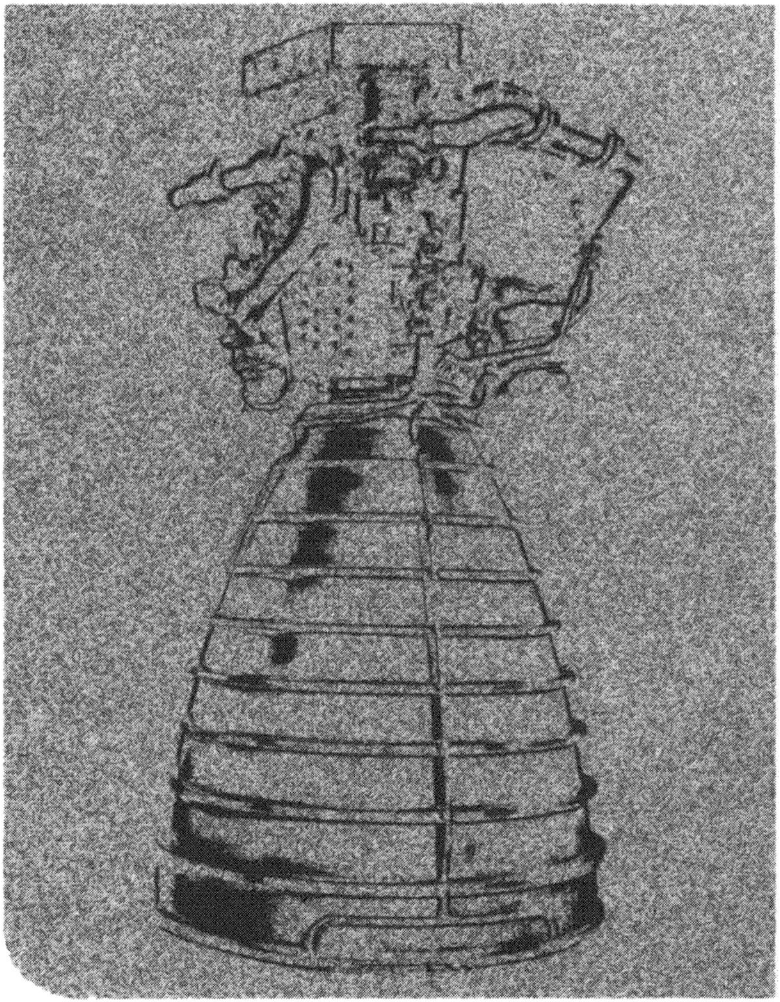

BC73-110

SPACE SHUTTLE MAIN ENGINE

DESIGN FEATURES

by

Rocketdyne Division
Rockwell International

ISBN # 978-1-937684-79-2
www.PeriscopeFilm.com

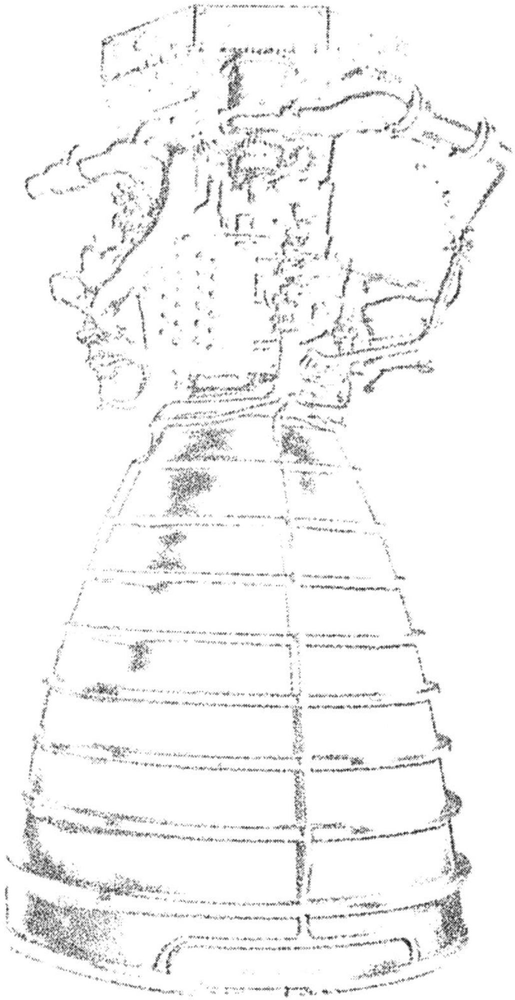

SPACE SHUTTLE MAIN ENGINE

*

DESIGN FEATURES

ISBN # 978-1-937684-79-2

www.PeriscopeFilm.com

INTRODUCTION

The material presented within this book is a description of the current design features of the Space Shuttle Main Engine and its major components.

The contents represent the latest information available on the SSME at this time, including illustrations of component parts as well as facts detailing operational characteristics.

September 1973

SPACE SHUTTLE

The space shuttle is a reusable transportation system that will substantially reduce the costs of earth orbital operations while improving operational capabilities and flexibility. The system will be capable of delivering payloads of up to 65,000 pounds and can return to earth with 40,000 pounds of payload, a capability unavailable with current expendable launch vehicles.

The space shuttle will safely and comfortably transport scientists, technicians, or astronauts into orbit while delivering payloads. This permits direct participation in space experiments and observations by top men and women in their fields, no longer limiting space flight to intensively trained, physically perfect astronauts.

Space shuttle missions will permit continuous updating of inventories of the earth's resources – water, crops, and minerals – allowing more effective application of these resources in meeting human needs. They will support the detection of air and water pollution sources, and they will permit improved weather predictions, resulting in the savings of billions of dollars in reduced property and crop damage.

Scientific payloads will acquire new data on the chemistry and physics of the sun and the stars, and may provide the key to unlocking the remaining secrets of fusion, facilitating the development of unlimited, pollution-free power for the needs of man. Advanced communications relay satellites can be used to bring educational programs to the people of under-developed countries and permit better understanding and trust among the nations of the world.

Improved navigation aids will enhance air and ocean travel safety and permit precision surveys of the earth's surface.

Advances in the biological and medical sciences accomplished by space-based research will prolong the span of useful, healthy life.

The space shuttle system will deliver many of the Department of Defense payloads to earth orbit. These missions will strongly deter war, rendering it impossible for a nation to covertly prepare for and launch a massive sneak attack on its neighbors. The chances of technological surprise and the secret development of weapons grossly altering the balance of power will be greatly reduced.

MISSION PROFILE

During launch, hold-down is provided until the main engines and the SRB's provide a thrust equal to the weight. Pitch and roll into the preferred attitude for the selected launch azimuth are initiated after the vehicle clears the launch tower approximately 5 seconds following liftoff. Maximum loads normal to the flight path can be expected about 60 seconds after liftoff for the due-east mission illustrated. Maximum dynamic pressure of approximately 650 pounds per square foot occurs at 39,400 feet.

Upon burnout, approximately 125 seconds after liftoff, the SRB's are separated. Small solid rocket motors force the empty cases away from the orbiter and external tank (ET), which continue toward orbit. The SRB's fall in an arc and are decelerated by drogue parachutes deployed at 16,000 feet. The SRB cases and recovery system are retrieved from the ocean and towed back to land for refurbishment and reuse.

Shortly before insertion into orbit, approximately 500 seconds after liftoff, the main engines are shut down and the ET is separated from the orbiter, resulting in atmospheric entry and impact in a preselected remote ocean area.

The orbiter is inserted into an elliptical orbit at a nominal perigee of 60 nautical miles altitude by the OMS.

At apogee, 100 nautical miles altitude, the orbit is modified to the one desired by using the OMS. Orbital operations involving payload deployment, observation, experiments, or other activities are then performed.

After completion of orbital operations, the OMS provides the velocity change necessary to perform the deorbit maneuver. The orbiter enters the atmosphere at a flight path angle of approximately 1 degree with an angle of attack of 34 degrees. A deceleration glide is then performed to reach the desired landing site. The orbiter can reach landing sites as far as 1085 nautical miles on either side of its initial flight path. After the orbiter glides into position, an unpowered landing is made.

The orbiter and its ground support system have been designed to permit turnaround of the vehicle for the next flight within 14 days after touchdown. This includes refurbishment, maintenance, assembly, and checkout prior to launch.

SPACE SHUTTLE MISSION PROFILE

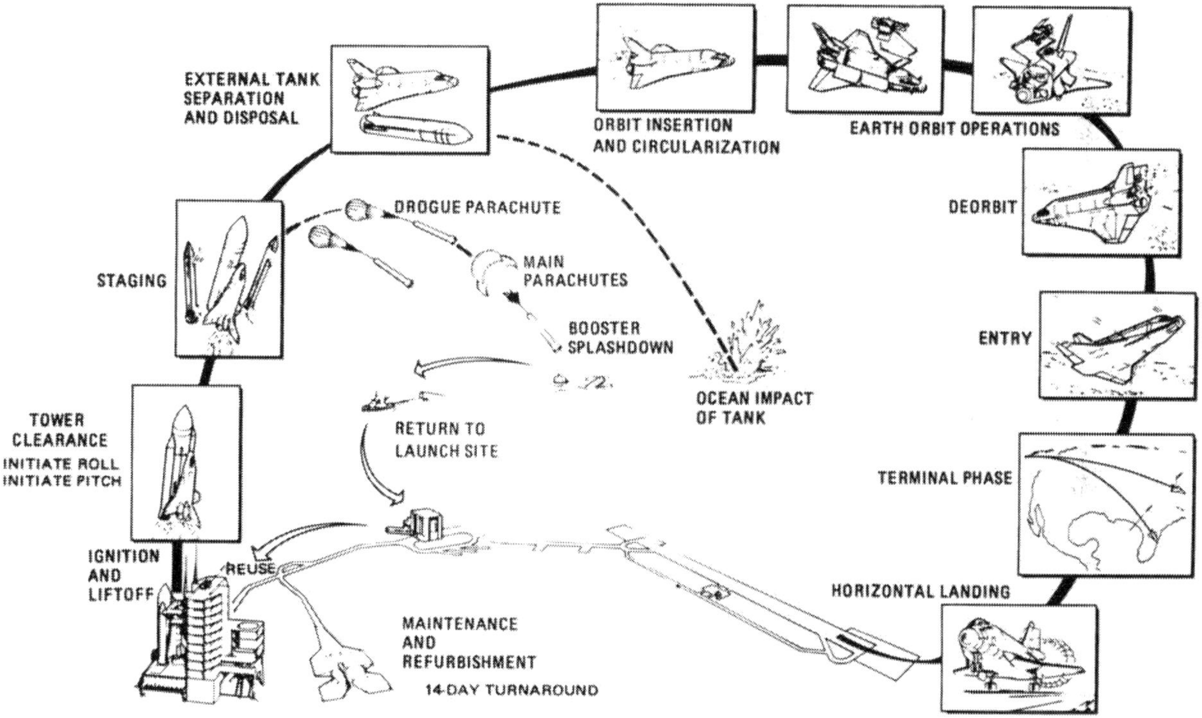

EXTERNAL TANK SEPARATION AND DISPOSAL

ORBIT INSERTION AND CIRCULARIZATION

EARTH ORBIT OPERATIONS

DEORBIT

ENTRY

TERMINAL PHASE

HORIZONTAL LANDING

STAGING

DROGUE PARACHUTE

MAIN PARACHUTES

BOOSTER SPLASHDOWN

OCEAN IMPACT OF TANK

RETURN TO LAUNCH SITE

TOWER CLEARANCE

INITIATE ROLL
INITIATE PITCH

IGNITION AND LIFTOFF

REUSE

MAINTENANCE AND REFURBISHMENT

14-DAY TURNAROUND

304-101

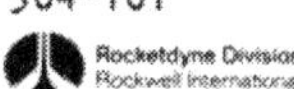

SHUTTLE CONFIGURATION

The integrated Space Shuttle vehicle consists of the orbiter, the external tank (ET), and the solid rocket boosters (SRB's). The orbiter carries the four-man crew and payload and is equipped with three 470,000-pound-thrust (vacuum) rocket engines. The external tank contains propellants for the main engines. The SRB's provide thrust augmentation during the initial phases of launch, up to a velocity of about 4470 feet per second. Combined sea level thrust of the main engines and the SRB's is 6.25 million pounds.

EXTERNAL TANK

The external tank contains all the propellants supplied to the orbiter main engines. These propellants consist of liquid hydrogen (LH_2) fuel and liquid oxygen (LO_2) oxidizer. All fluid controls and valves for main propulsion system (MPS) operation are located in the orbiter to reduce recurring costs. Antivortex and slosh baffles are mounted in the oxidizer tank to minimize liquid residuals and damp fluid motion. Five lines (three fuel and two oxidizer) interface between the external tank and the orbiter. All interface lines except the oxidizer vent line are insulated with spray-on foam and protected with a fiberglass fairing. An uninsulated antigeyser line on the external tank provides LO_2 geyser suppression. Liquid-level point sensors are used in both tanks for loading control. The external tank contains 1.55 million pounds of usable propellant at liftoff. The liquid hydrogen tank volume is 53,800 cubic feet, and the liquid oxygen tank volume is 19,500 cubic feet. These volumes include a 3-percent ullage provision. The hydrogen and oxygen tanks are pressurized to respective ranges of 32 to 34 psia and 20 to 22 psia. Both tanks will be constructed of aluminum alloy skins with support or stability frames as required. The primary structural attachment to the orbiter consists of one forward and two rear connections. Spray-on foam insulation (SOFI) is applied to the complete outer surface of the LH_2 tank. An ablator of sheet cork is bonded directly to the outer surface of the LO_2 tank nose cone and areas of the intertank skirt structure adjacent to the SRB's.

SOLID ROCKET BOOSTERS

Two SRB's are attached to the external tank and burn in parallel with the main propulsion system, providing ascent propulsive thrust up to staging. The primary elements of the motors are the case, nozzle and thrust vector control (TVC), propellant system, igniter with safe and arm provisions, and thrust termination and malfunction detection instrumentation subsystems. Each SRB weighs approximately 1.1 million pounds and produces 2.56 million pounds of thrust at sea level. The propellant grain is shaped to reduce thrust and prevent overstressing the vehicle after liftoff. The grain is of conventional design, employing a star perforation in the forward motor closure and a truncated cone perforation in each of the segments and aft closure. The contoured nozzle expansion ratio (area of exit to area of throat) is 11 to 1. The SRB TVC gimbals the nozzle ±5 degrees, with a 1-degree override capability. The SRB's are released by pyrotechnic separation devices. Thrusters on each SRB, at the aft and forward ends, provide separation from the orbiter/tank.

THE SPACE SHUTTLE

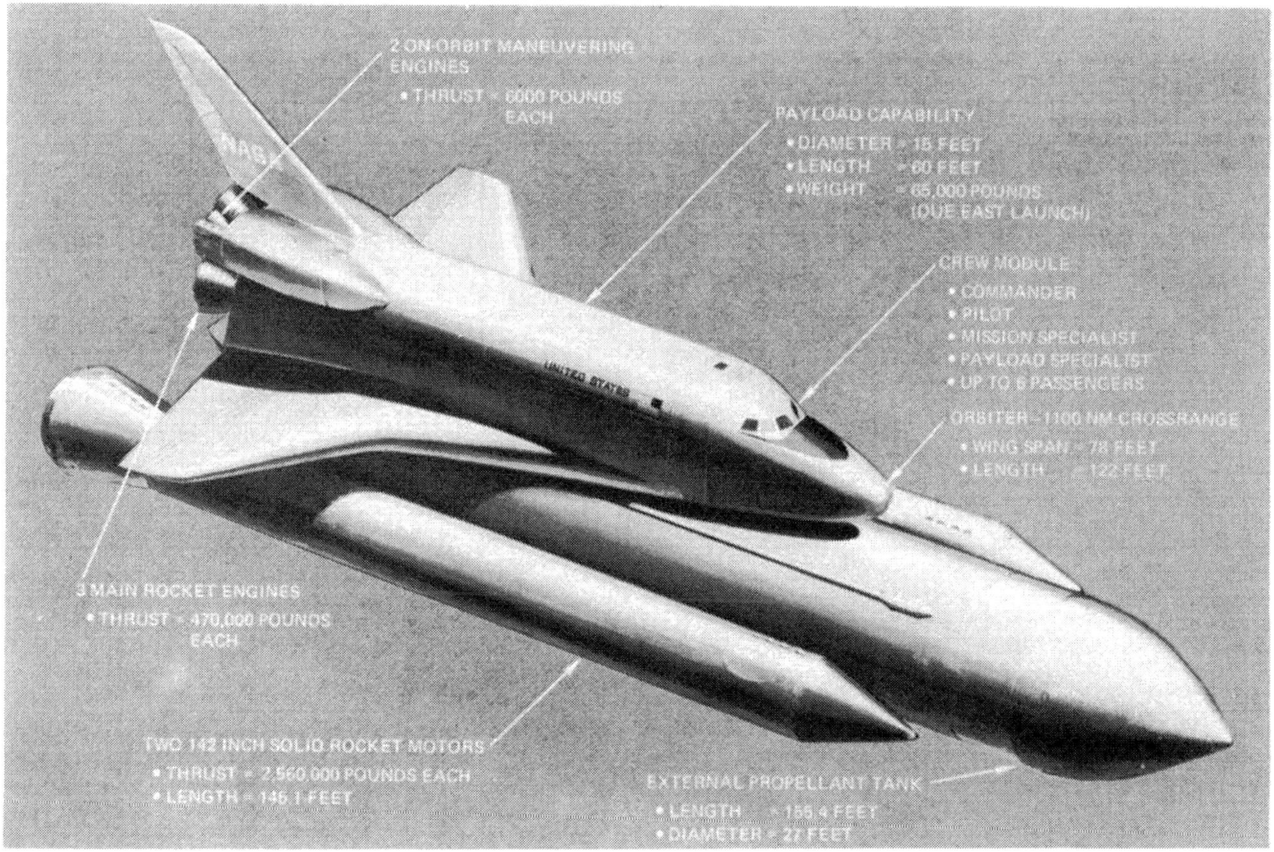

2 ON-ORBIT MANEUVERING ENGINES
• THRUST = 6000 POUNDS EACH

PAYLOAD CAPABILITY
• DIAMETER = 15 FEET
• LENGTH = 60 FEET
• WEIGHT = 65,000 POUNDS (DUE EAST LAUNCH)

CREW MODULE
• COMMANDER
• PILOT
• MISSION SPECIALIST
• PAYLOAD SPECIALIST
• UP TO 6 PASSENGERS

ORBITER–1100 NM CROSSRANGE
• WING SPAN = 78 FEET
• LENGTH = 122 FEET

3 MAIN ROCKET ENGINES
• THRUST = 470,000 POUNDS EACH

TWO 142 INCH SOLID ROCKET MOTORS
• THRUST = 2,560,000 POUNDS EACH
• LENGTH = 146.1 FEET

EXTERNAL PROPELLANT TANK
• LENGTH = 155.4 FEET
• DIAMETER = 27 FEET

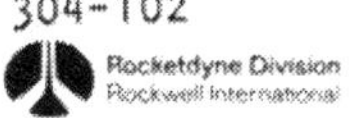

SPACE SHUTTLE VEHICLE

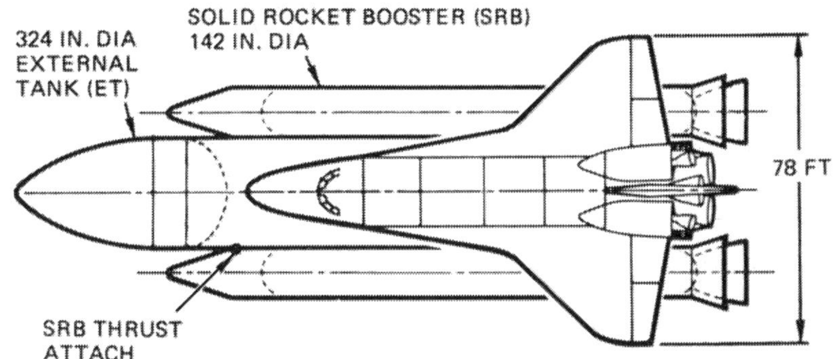

GROSS LIFT-OFF WEIGHT	4203K LB
(50 X 100 N MILES BY 28.5 DEGREES)	

SRB	2327K LB
ET	1631K LB
ORBITER	
DRY	150K LB
CREW AND PROPELLANTS	30K LB
PAYLOAD	65K LB

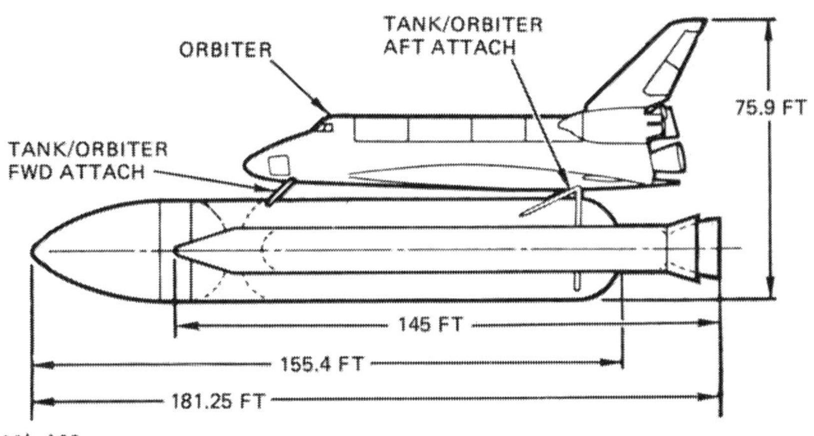

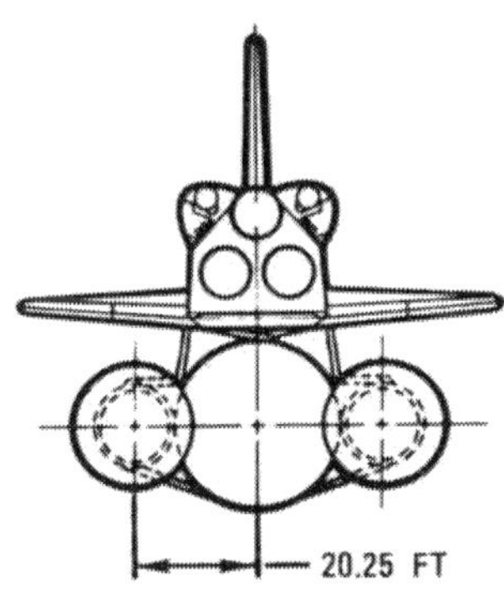

304-103

ORBITER VEHICLE REQUIREMENTS

Crew Accommodations

- Basic Four-Man Crew
- 7-Day Mission (28-Man Day Support)
- Six Additional Crewmen
- Short-Duration Mission (42-Man Day Support)
- Design Not To Preclude 30-Day Mission
- Extra-Vehicular Activity/Intra-Vehicular Activity

Land 25,000-Pound Payload

Docking Provisions (Daylight Or Dark)

Cooperative Target In-Plane Rendezvous (300-Nautical Mile Displacement)

Rendezvous And Retrieve Passive Stabilized Element

Operate On Runways 150 By 10,000 Feet

Normal Flight Control And Structural Dynamics

- Space And Atmospheric Operations
- Subsonic And Hypersonic Flight

Minimum Environmental Impact

Minimize Payload Contamination (RCS Impingement, Etc.)

304-104T

SPACE SHUTTLE ORBITER

304-104

The orbiter, comparable in size and weight to modern transport aircraft-size of DC-9, contains the crew and payload for the Space Shuttle system. The orbiter can deliver single or multiple payloads of up to 65,000 pounds; the payloads can be up to 60 feet long and have a diameter of up to 15 feet. The orbiter crew compartment can accommodate up to 10 persons, including the crew. Aerodynamic control surfaces augment control of the integrated Space Shuttle vehicle provided by the gimbaled main engines during boost and provide control of the orbiter below Mach 2. Design touchdown speed is 165 knots, consistent with current high-performance aircraft.

MAIN PROPULSION SYSTEM (MPS)

The MPS, consisting of three Space Shuttle Main Engines (SSME's), operates in parallel with the SRB's during the initial ascent phase and continues to burn until just before injection after SRB separation. Each of the rocket engines operate at a mixture ratio (liquid oxygen/liquid hydrogen) of 6:1 and a chamber pressure of approximately 3000 psia to produce a sea level thrust of 375,000 pounds and a vacuum thrust of 470,000 pounds with a fixed nozzle area ratio of 77.5:1. The engines are throttleable over a thrust range of 50 to 109 percent of the design thrust level. This provides a higher thrust level during liftoff and the initial ascent phase, and allows limiting orbiter acceleration to 3 g's during the final ascent phase. The engines are gimbaled to deflect ± 11 degrees for pitch, yaw, and roll control during the orbiter boost phase.

ORBITAL MANEUVERING SYSTEM

The orbital maneuvering system (OMS), consisting of two liquid propellant rocket engines, provides the propulsive thrust to perform orbit injection, orbit circularization, orbit transfer, rendezvous, and deorbit. The engines are mounted in two pods, one located on each side of the aft fuselage. Each pod contains a high-pressure helium storage bottle, tank pressurization regulators and controls, a fuel tank, an oxidizer tank, and a pressure-fed, regeneratively cooled, rocket engine. Each engine produces a vacuum thrust of 6000 pounds and uses nitrogen tetroxide (N_2O_4) as the oxidizer and monomethylhydrazine (MMH) as the fuel.

REACTION CONTROL SYSTEM

The orbiter reaction control system (RCS) will provide vehicle attitude control in space and translation capability for small velocity increments. These functions will be provided for just prior to separation from the external main propellant tank through the on-orbit maneuvers and during certain phases of entry. The RCS consists of 46 bipropellant (H_2O_4/MMH) thrusters to provide fail operational/fail safe attitude control and translation capability. Forty of the thrusters are 900 pound thrust units and six are 25 pound vernier thrust units. The 6 vernier units and 16 of the 900 pound thrusters are contained in a forward module and the other 24 are divided equally between two rear modules. Each module contains propellant tanks which use a positive expulsion device to ensure propellant feed under all operating conditions. The proposed configuration is sized to meet the propellant requirements of a polar mission.

304-105T

ORBITER VEHICLE
78 FT
2 OMS/RCS PODS
• OMS THRUST = 6K/EA
• RCS THRUST = 0.90K/EA
123 FT
26.3% L_B
82.8% L_B
FWD RCS
57 FT
3 SSME
470K THRUST/EA
REF BODY LENGTH (L_B) = 107.5 FT
238
1528

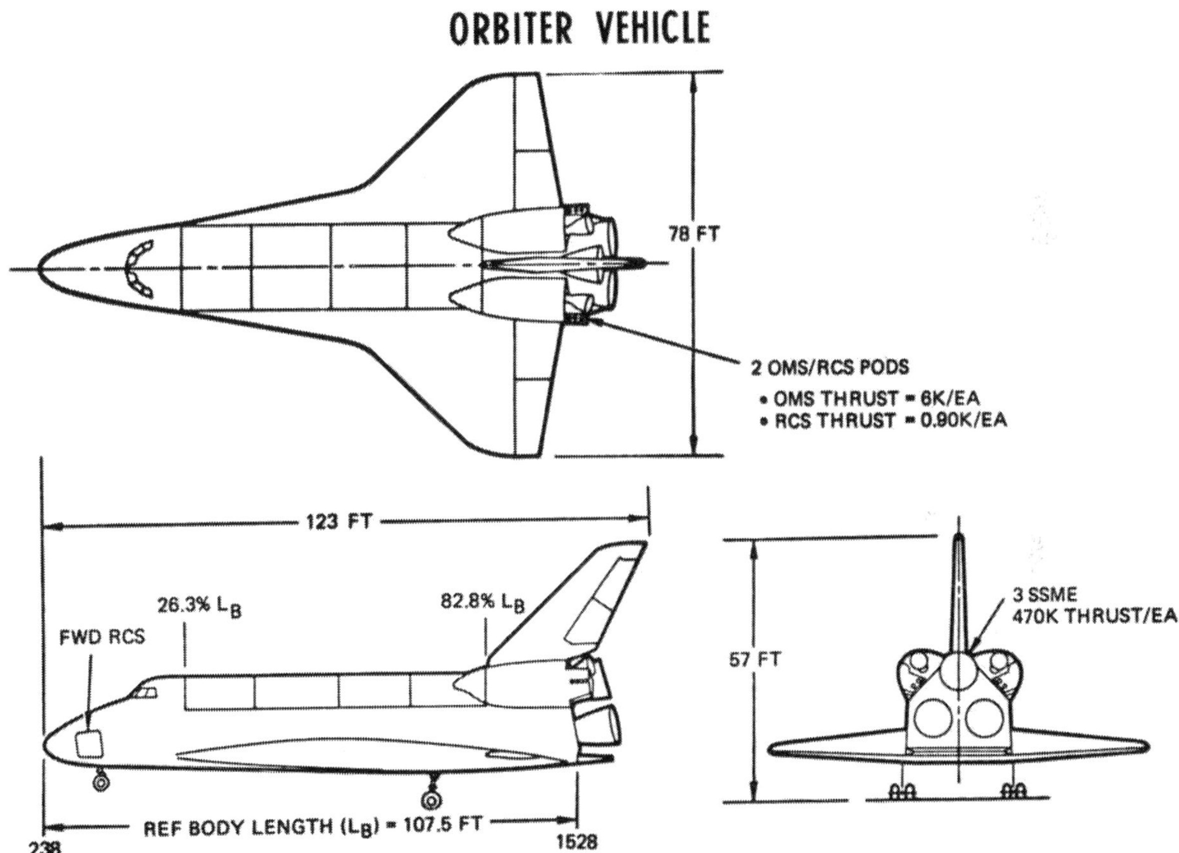

304-105

SPACE SHUTTLE MAIN ENGINE

The SSME is a reusable, high-performance, liquid-propellant rocket engine with a variable thrust and mixture ratio. It is being developed to provide the primary thrust for the National Aeronautics and Space Administration's Space Shuttle orbiter vehicles. Three of the engines are clustered on each orbiter vehicle. They are ignited on the ground at launch and burn for an average of 10 minutes during vehicle boost to earth orbit.

Many unique and innovative features are being designed into the SSME to satisfy the performance, life, reliability, and maintainability requirements of the Space Shuttle vehicle.

Significant to meeting performance requirements is the use of a staged combustion power cycle coupled with high combustion chamber pressures. In the SSME staged combustion cycle, the propellants are partially burned at high pressure and relatively low temperature in the preburners, then completely combusted at high temperature and pressure in the main chamber before expanding through the high-area-ratio nozzle.

Using hydrogen fuel to cool all combustion devices directly exposed to contact with high-temperature combustion products contributes to long life.

An electronic engine controller automatically performs checkout, start, mainstage, and engine shutdown functions ensuring high reliability.

Easy accessibility to line replaceable units (LRU's) and the use of internal inspection ports in critical components aid maintainability.

304-106T

SPACE SHUTTLE MAIN ENGINE

• THRUST	
• SEA LEVEL	375,000 lbf
• VACUUM	470,000 lbf
• EPL	109%
• CHAMBER PRESSURE	2970 PSIA
• AREA RATIO	77.5
• SPECIFIC IMPULSE (NOM)	
• SEA LEVEL	363.2
• VACUUM	455.2
• MIXTURE RATIO	6.0
• LENGTH	167"
• DIAMETER	
• POWERHEAD	105" × 94.5"
• NOZZLE EXIT	94"
• LIFE	7.5 HRS 100 STARTS: 6 EPL'S 94 NPL'S

SSME PHYSICAL CHARACTERISTICS

Four turbopumps, two low-pressure and two high-pressure, are key components in describing the physical characteristics of the SSME system. The low-pressure fuel turbopump (LPFTP) and the low-pressure oxidizer turbopump (LPOTP) are located 180 degrees apart. The high-pressure fuel turbopump (HPFTP) and high-pressure oxidizer turbopump (HPOTP) also located 180 degrees apart are mounted between the two low-pressure pumps. The LPFTP and LPOTP are connected to the vehicle propellant ducting and are supported in a fixed position by the vehicle structure. The discharge of each low-pressure pump is connected to the inlet of the high-pressure pump by ducts that have bellows systems to permit deflection required during gimbaling. The low-pressure pumps are axial flow-type pumps and operate at low speed to provide sufficient pressure to eliminate cavitation at the inlets of the high-pressure pumps.

The HPFTP is a three-stage, centrifugal flow-type pump, driven by a two-stage hot-gas turbine and is flange-attached to the hot gas manifold (HGM). It supplies liquid hydrogen to the main chamber and nozzle coolant circuits.

The HPOTP is flange-attached to the HGM and consists of two centrifugal-type pumps on a common shaft driven by a two-stage hot-gas turbine. The main pump supplies oxidizer to the main chamber injector and preburner pump (the second pump of the HPOTP). The preburner oxidizer pump raises the pressure of the oxidizer supplied to the fuel and oxidizer preburners.

The HGM is the structural backbone of the engine package in that it supports the two preburners, high-pressure pumps, main injector, and main chamber. It interconnects the fuel and oxidizer preburners (FPB and OPB) to the main chamber injector.

Both FPB and OPB are welded to the HGM and generate the fuel-rich gases that power the HPFTP and HPOTP. Each preburner consists of a combustor with a single-pass, fuel-cooled jacket and a baffled, coaxial-element injector.

The main injector and dome assembly is welded to the HGM. It is a baffled, coaxial-element injector, having dual faceplates that are transpiration-cooled by gaseous hydrogen. The gimbal bearing is bolted to the main injector and dome assembly and is the thrust interface between the engine and vehicle. It enables the engine to be gimbaled for thrust vector control.

The main combustion chamber (MCC) is bolted to the HGM and consists of an internal coolant liner and an external structural jacket. The coolant liner has longitudinal slots to form a fuel-cooled, uppass cooling circuit.

The nozzle is bolted to the MCC and is constructed of tapered tubes reinforced with insulated hat bands. The tubes provide a fuel-cooled, uppass cooling circuit that is supplied with hydrogen through three downpass transfer ducts connecting the nozzle forward manifold to the aft inlet manifold.

The controller is attached to the MCC by three clevis-type fittings. It is a solid-state, integral electronics package contained in a pressurized aluminum case assembly. It sequences, controls, and monitors all engine functions during checkout and operation of the engine.

304-107T

SSME MAJOR COMPONENTS

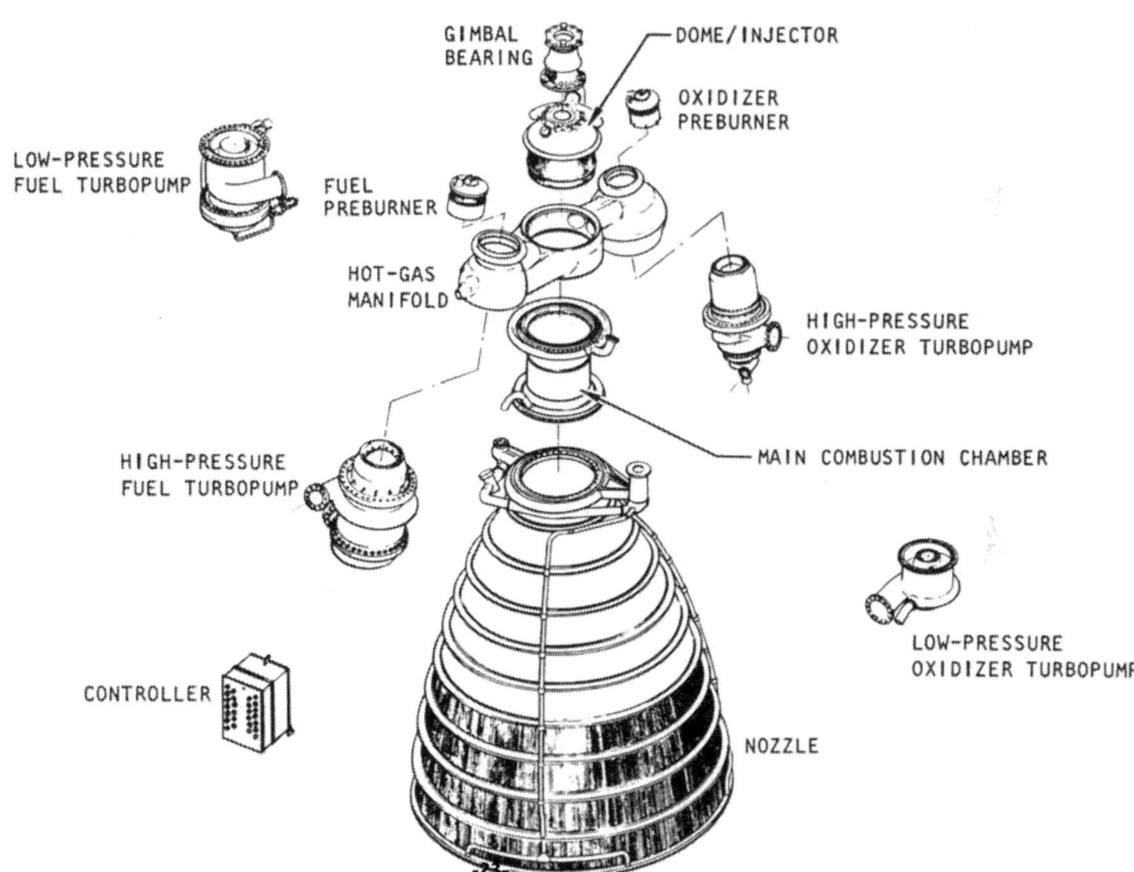

304-107

Rocketdyne Division
Rockwell International

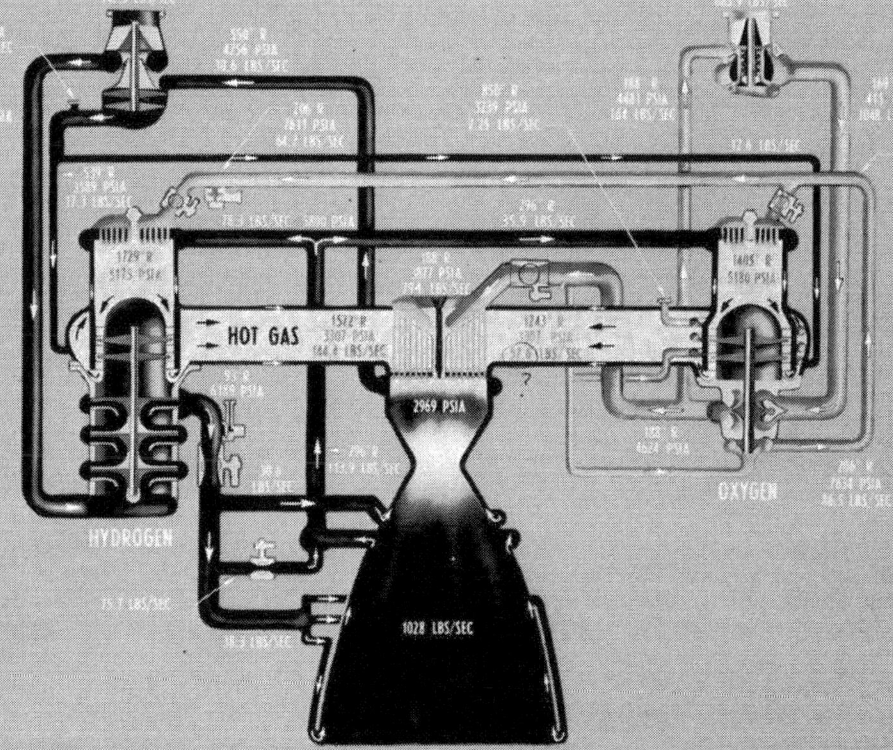
SSME PROPELLANT FLOW SCHEMATIC
HOT GAS
2969 PSIA
1028 LBS/SEC
HYDROGEN
OXYGEN
LC300-267H
Rocketdyne Division
Rockwell International

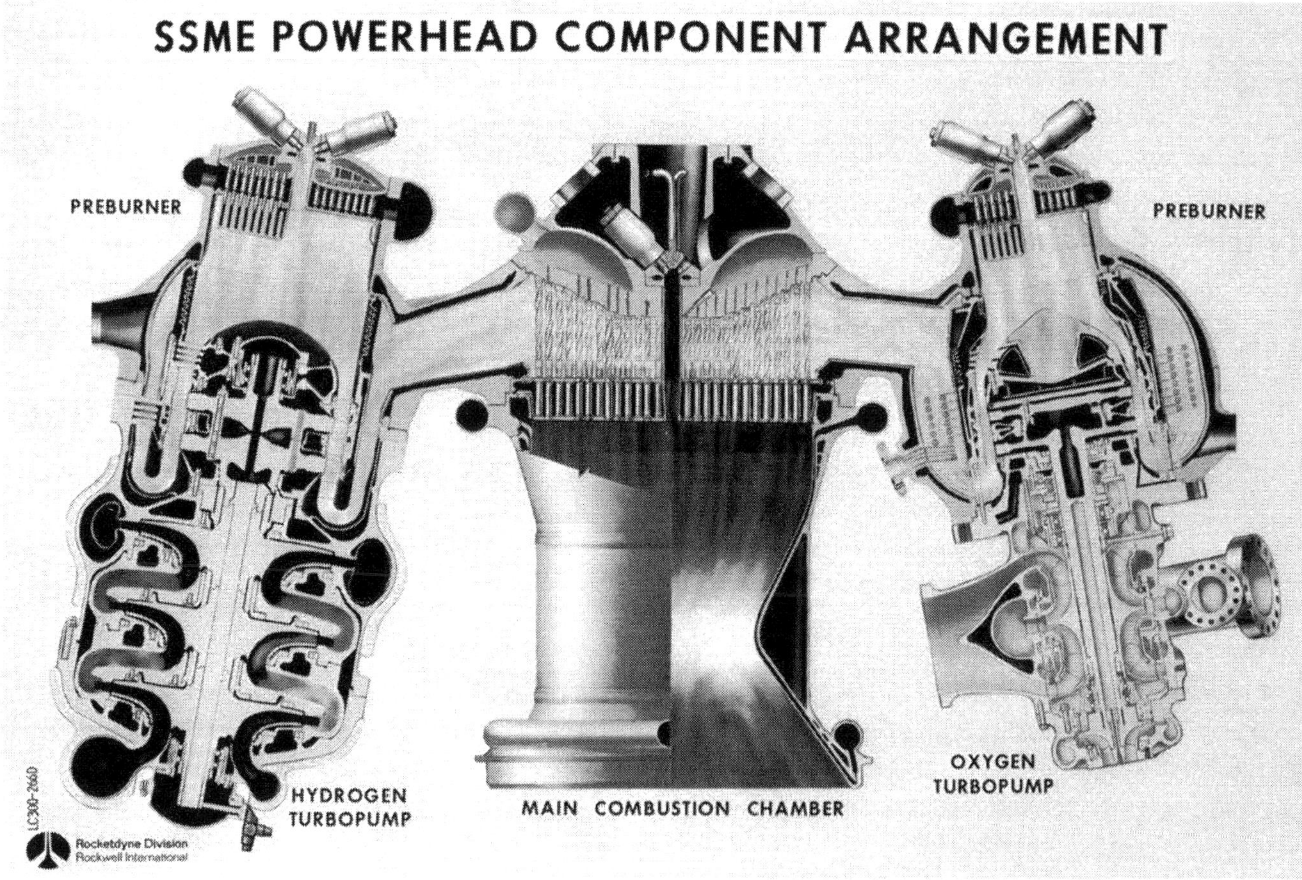

SSME POWERHEAD COMPONENT ARRANGEMENT

PREBURNER

PREBURNER

HYDROGEN
TURBOPUMP

MAIN COMBUSTION CHAMBER

OXYGEN
TURBOPUMP

LC300-2660

Rocketdyne Division
Rockwell International

HOT-GAS MANIFOLD

The hot-gas manifold (HGM) is a double-walled, hydrogen-gas-cooled structural support and fluid manifold. It is the structural backbone of the engine and interconnects and supports the preburners, high-pressure turbopumps, main combustion chamber, and main injector.

The HGM conducts hot gas from the turbines to the main chamber injector. The HGM is a machined Inconel 718 forging insulated with an inner liner. The liner consists of layers of 316L CRES perforated foil sheets alternated with layers of 316L CRES wire screen, diffusion bonded to Haynes 188 liner. The HGM assembly weighs approximately 600 pounds. The area between the wall and liner provides a coolant flow path for the hydrogen gas that exhausts from the low-pressure fuel turbine. This protects the outer wall and liner against the temperature effects of the hot gas from the preburners. After cooling the manifold, the coolant is directed between the two main injector Rigimesh faceplates.

The oxidizer side of the HGM has a canted flange to which the high-pressure oxidizer turbopump (HPOTP) is bolted. Heat exchanger tube supports are welded to the inner wall on the oxidizer side of the HGM. Two hot-gas transfer tubes route the HPOTP turbine exhaust gas to the main injector torus manifold, where it is radially directed into the hot-gas cavity of the main injector. The oxidizer preburner is welded to the upper end of the oxidizer side of the HGM.

The fuel side of the HGM also has a canted flange, to which the high-pressure fuel turbopump (HPFTP) is bolted. Three hot-gas transfer tubes route the HPFTP turbine exhaust gas to the main injector torus manifold where it is radially directed into the hot-gas cavity of the main injector. The fuel preburner is welded to the upper end of the fuel side of the HGM.

Instrumentation ports, which are also used for internal inspection of the injectors and turbines, are located throughout the HGM.

Geometry

- Structural Interface Between Seven Major Components
- Serves As Major Propellant Duct
- INCO 718 Outer Structural Shell Plus Liner Of Haynes 188 And 316L CRES Laminate
- Hydrogen-Cooled Shell

Operating Parameters (NPL MR = 6.0)

	Hot Gas	Coolant
Pressure (Maximum), psia	3361	3741
Temperature (Maximum), R	1573	540
Flowrate (Total), lb/sec	202	29.9

304-108T

SSME HOT GAS MANIFOLD

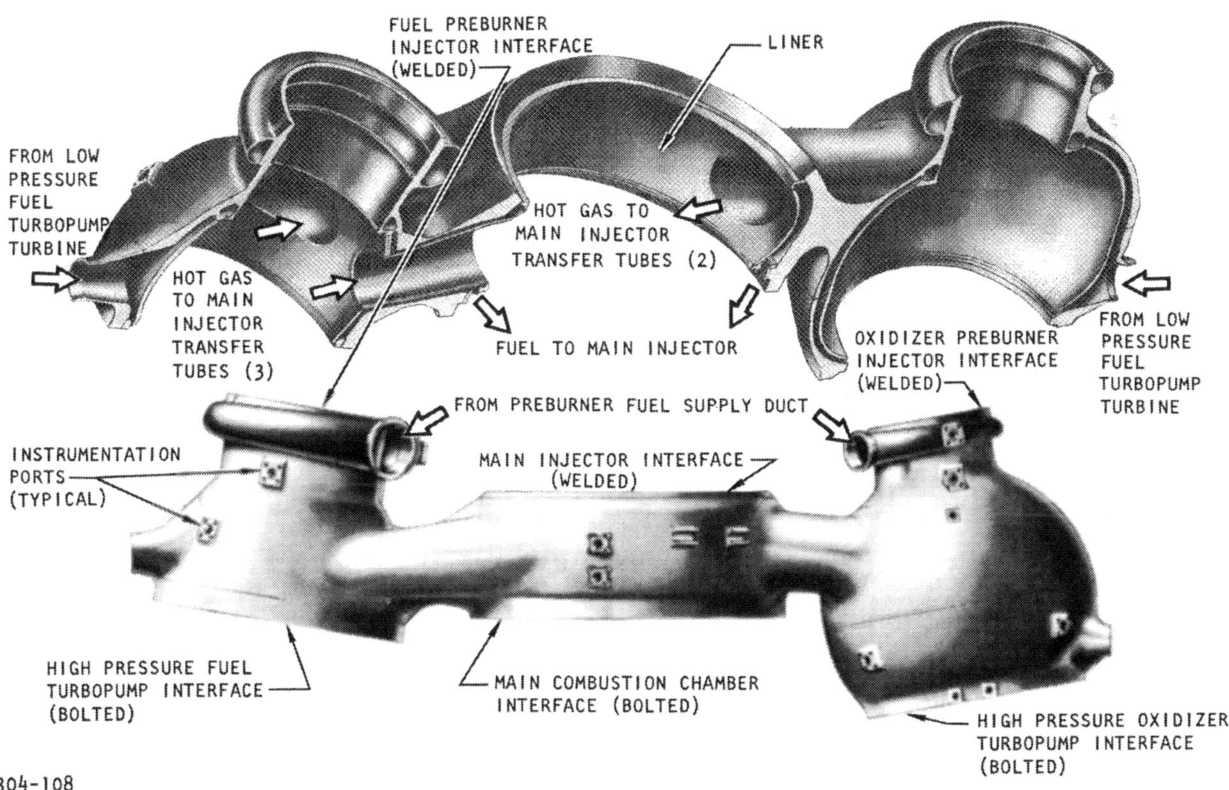

304-108

Rocketdyne Division
Rockwell International

PREBURNERS

Two preburners are used in the SSME. They burn hydrogen and oxygen to generate a variable hot-gas supply to power the turbopumps. They operate at a low mixture ratio with gaseous hydrogen from the main chamber nozzle coolant circuit and liquid oxygen from the preburner oxidizer pump. Specific operating levels of the preburners are controlled by regulating the oxidizer flowrate with the preburner oxidizer valve.

The preburner injector is a baffled, coaxial-element injector that mixes the gaseous hydrogen and liquid oxidizer in correct proportions, and uniformly distributes and injects the propellants into the combustion chamber. The coaxial elements are contained by the upper interpropellant plate and a lower faceplate in a closely spaced pattern of concentric rows. Each element consists of an orificed center tube (oxidizer post) and an outer fuel sleeve that has a series of slots cut in its periphery. Oxidizer entering the oxidizer manifold is uniformly distributed within the oxidizer dome over the interpropellant plate and enters the oxidizer post through the post orifice. Hydrogen from the fuel manifold passes radially into the injector fuel cavity formed by the interpropellant plate and the faceplate. From the fuel cavity, the hydrogen enters the annulus of the coaxial elements through the fuel sleeve slots. The high injection velocity of the fuel, relative to the velocity of the oxidizer, produces a high rate of atomization and thorough mixing. Three NARloy-A baffles partition the injector face to aid in stable combustion. Hydrogen flows through axial passages in each baffle for cooling and is discharged radially into the combustion chamber.

The ASI chamber, a small combustion chamber in the center of the injector, initiates ignition of the FPB propellants. Integral to the oxidizer dome, an impinging doublet oxidizer injector and a fuel injector with eight tangential orifices supply propellants to the ASI combustion chamber where they are ignited by dual redundant spark igniters.

304-109T

PREBURNER INJECTION ELEMENT

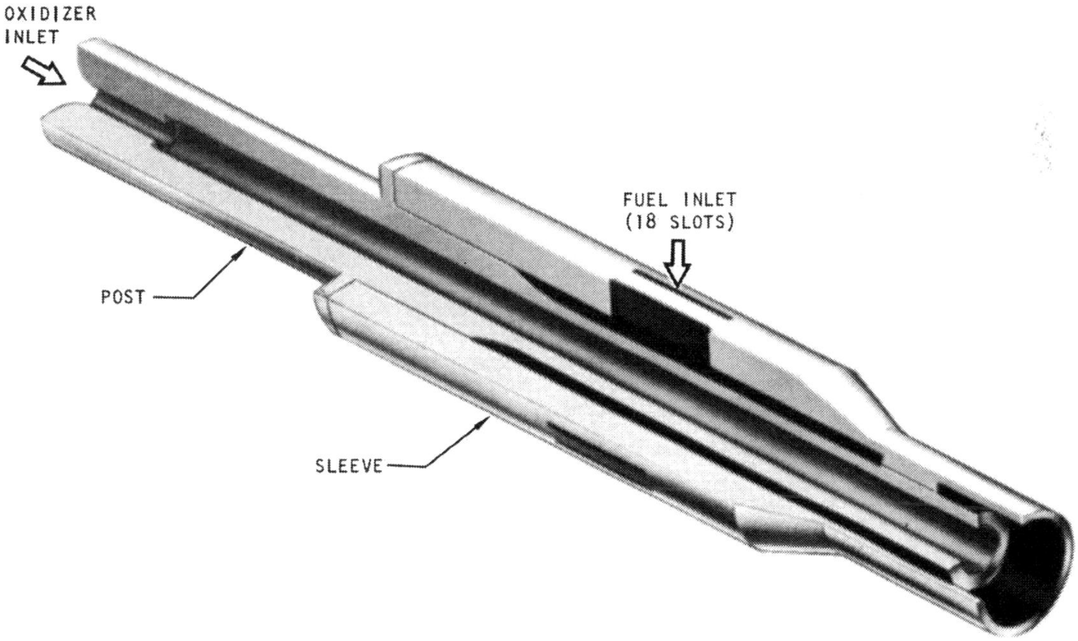

OXIDIZER INLET

FUEL INLET (18 SLOTS)

POST

SLEEVE

304-109

FUEL PREBURNER

Weighing approximately 135 pounds, the fuel preburner (FPB) is supported by the hot-gas manifold (HGM) to which it is welded, and is close-coupled to the high-pressure fuel turbopump (HPFTP). The FPB consists of three major parts: (1) injector, (2) augmented spark igniter (ASI) chamber, and (3) combustion chamber.

The FPB combustion chamber is a fuel-cooled, double-walled, 10-inch-ID chamber in which energy is generated to power the HPFTP. Consisting of a nickel-base-alloy outer wall and a Haynes 188 liner with nine acoustic absorbers, the chamber is welded to the injector/dome and HGM. Fuel coolant is provided between the outer wall and liner by hydrogen supplied from the fuel manifold. The fuel coolant is discharged at the lower end of the liner into the hot-gas powering the HPFTP turbine.

Geometry	
Internal Diameter, Inches	10.43
Combustor Length, Inches	4.37
Faceplate Material	Inconel 625
Injector Configuration	Concentric Orifice
Number of Elements	264
Baffle Length, Inches	2.25
Baffle Material	NARloy-A
Operating Parameters (NPL, MR = 6.0)	
Injector-End Pressure, psia	5176
Combustion Temperature, R	1729
Hot-Gas Mixture Ratio (o/f)	0.829
Oxidizer Flowrate (including Igniter), lb/sec	64.7
Fuel Flowrate (including Igniter), lb/sec	78.0

304-110T

FUEL PREBURNER

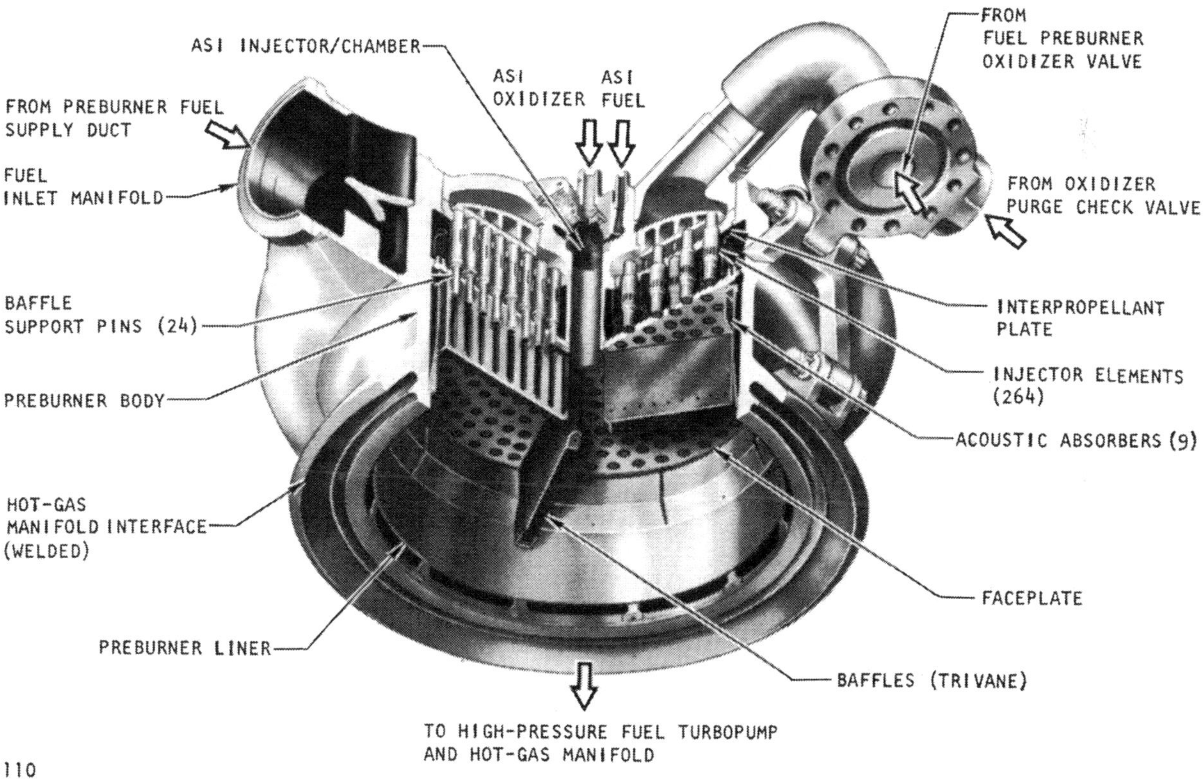

ASI INJECTOR/CHAMBER

ASI OXIDIZER

ASI FUEL

FROM FUEL PREBURNER OXIDIZER VALVE

FROM PREBURNER FUEL SUPPLY DUCT

FUEL INLET MANIFOLD

FROM OXIDIZER PURGE CHECK VALVE

BAFFLE SUPPORT PINS (24)

INTERPROPELLANT PLATE

PREBURNER BODY

INJECTOR ELEMENTS (264)

ACOUSTIC ABSORBERS (9)

HOT-GAS MANIFOLD INTERFACE (WELDED)

FACEPLATE

PREBURNER LINER

BAFFLES (TRIVANE)

TO HIGH-PRESSURE FUEL TURBOPUMP AND HOT-GAS MANIFOLD

OXIDIZER PREBURNER

The OPB combustion chamber is a fuel-cooled, double-walled, 7.5-inch-ID chamber in which the energy is generated to power the HPOTP. Consisting of a nickel-base-alloy outer wall and a Haynes 188 liner with nine acoustic absorbers, the chamber is welded to the injector/dome and HGM. Fuel coolant is provided between the outer wall and liner by hydrogen supplied from the fuel manifold. The fuel coolant is discharged at the lower end of the liner into the hot-gas powering the HPOTP turbine.

Weighing approximately 80 pounds, the oxidizer preburner (OPB) is supported by the hot-gas manifold (HGM) to which it is welded, and is close-coupled to the high-pressure oxidizer turbopump (HPOTP). The OPB consists of three major parts: (1) injector, (2) augmented spark igniter (ASI) chamber, and (3) combustion chamber.

Geometry	
Internal Diameter, Inches	7.43
Combustor Length, Inches	4.25
Faceplate Material	Inconel 625
Injector Configuration	Concentric Orifice
Number Of Elements	120
Baffle Length, Inches	2.25
Baffle Material	NARloy-A
Operation Parameters (NPL, MR = 6.0)	
Injector-End Pressure, psia	5181
Combustion Temperature, R	1406
Hot-Gas Mixture Ratio (o/f)	0.633
Oxidizer Flowrate (Including Igniter), lb/sec	21.8
Fuel Flowrate (Including Igniter), lb/sec	36.0

304-111T

OXIDIZER PREBURNER

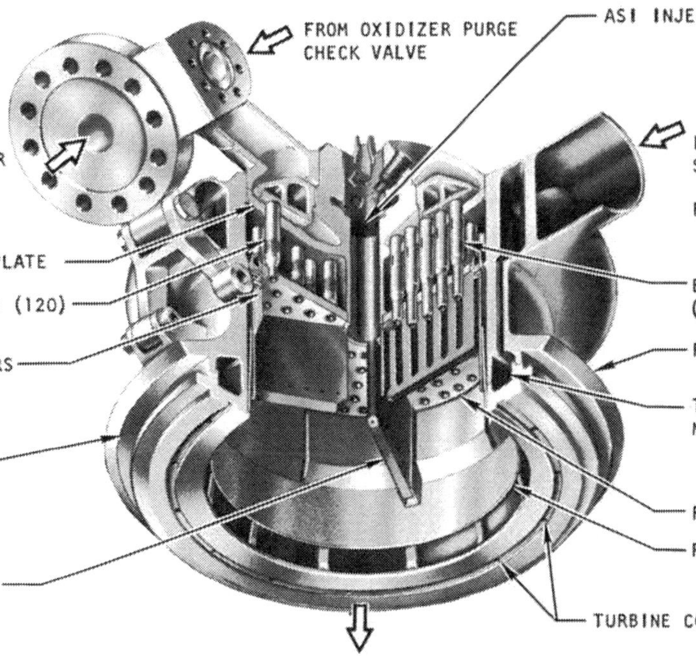

FROM OXIDIZER PURGE CHECK VALVE

ASI INJECTOR/CHAMBER

FROM OXIDIZER PREBURNER OXIDIZER VALVE

FROM PREBURNER FUEL SUPPLY DUCT

FUEL INLET MANIFOLD

INTERPROPELLANT PLATE

INJECTOR ELEMENTS (120)

BAFFLE SUPPORT PINS (15)

ACOUSTIC ABSORBERS

PREBURNER BODY

TURBINE COOLANT MANIFOLD

HOT-GAS MANIFOLD INTERFACE (WELDED)

FACEPLATE

PREBURNER LINER

BAFFLES (TRIVANE)

TURBINE COOLANT ORIFICES (36)

TO HIGH-PRESSURE OXIDIZER TURBOPUMP AND HOT-GAS MANIFOLD

304-111

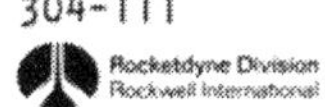

MAIN INJECTOR

The main injector is a baffled, coaxial element-type injector that efficiently mixes and uniformly distributes propellants to the main combustion chamber (MCC). It is approximately 22 inches in diameter, 19 inches long, and 380 pounds in weight. It is supported by the hot-gas manifold, to which it is welded, and interfaces with the MCC at a metal (Haynes 25) contracting-ring seal at the primary faceplate of the injector. A gimbal bearing mounts to the forward end of the injector and transmits thrust loads to the vehicle thrust structure. Basically, the injector assembly consists of an INCO 718 structural body, injection elements, two Rigimesh faceplates, and the ASI assembly.

The structural body contains the thrust cone, to which the gimbal bearing bolts; the injector body, which is welded to the thrust cone and to which the injector elements are welded; and the ASI assembly, which is welded to the injector body. The thrust cone and injector body form the oxidizer manifold.

The coaxial injection elements consist of 525 main injection elements and 75 baffle elements. The main and baffle injection elements are supplied with oxidizer from the injector oxidizer manifold. The main injection elements are supplied with fuel-rich hot gas from the turbine exhaust of the high-pressure turbopumps through the HGM. The baffle elements are supplied with gaseous hydrogen used in the HGM coolant circuit.

Both primary and secondary Rigimesh faceplates are supported by the injector elements and form the cavity from which part of the HGM cooling circuit hydrogen is supplied to the baffle elements. The primary faceplate separates the cooling hydrogen from the combustion chamber and the secondary faceplate separates the cooling hydrogen from the hot gas that is supplied as fuel to the main injector elements. Both faceplates are porous and are transpiration cooled by part of the HGM cooling circuit hydrogen.

The ASI assembly is welded to the center of the injector body and consists of a combustion chamber integral with oxidizer and fuel ASI injectors.

304-112T

MAIN INJECTOR ASSEMBLY

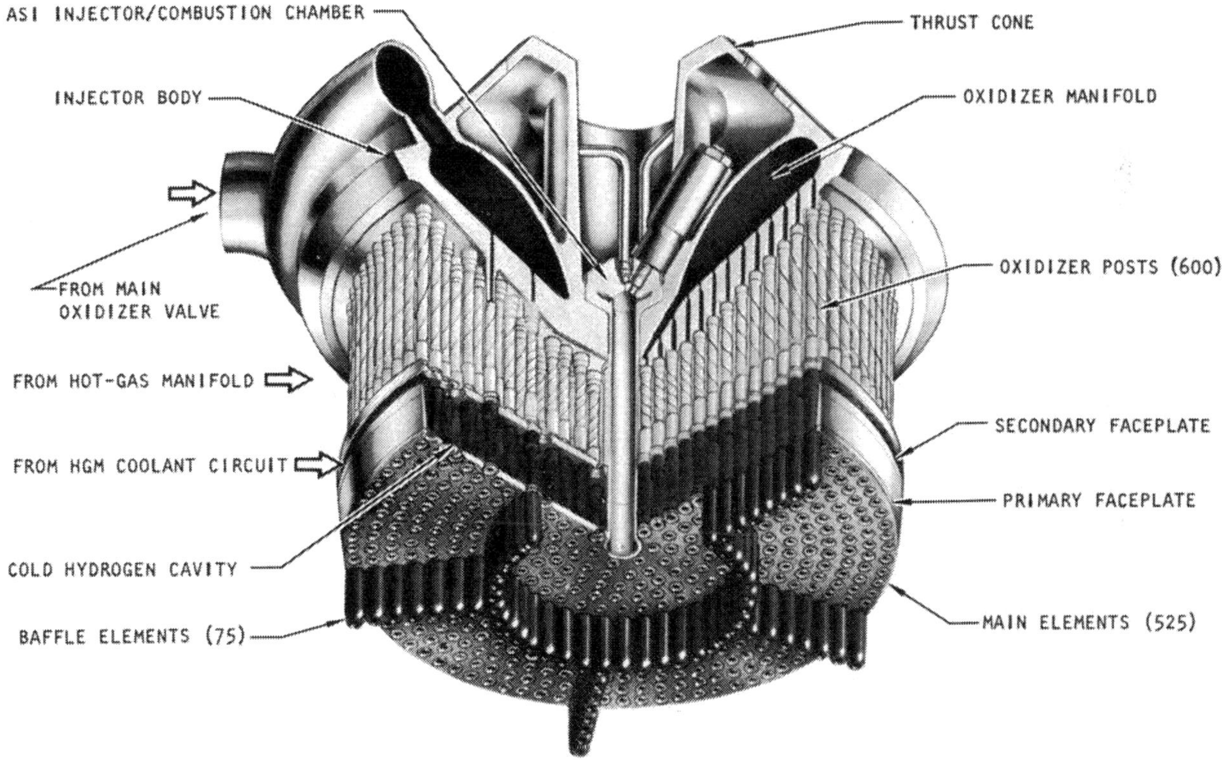

ASI INJECTOR/COMBUSTION CHAMBER

THRUST CONE

INJECTOR BODY

OXIDIZER MANIFOLD

FROM MAIN OXIDIZER VALVE

OXIDIZER POSTS (600)

FROM HOT-GAS MANIFOLD

SECONDARY FACEPLATE

FROM HGM COOLANT CIRCUIT

PRIMARY FACEPLATE

COLD HYDROGEN CAVITY

BAFFLE ELEMENTS (75)

MAIN ELEMENTS (525)

304-112

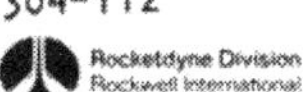

MAIN INJECTOR ELEMENTS

The main injector elements are bipropellant, gas/liquid, coaxial fluid-injection devices. They meter, mix, and distribute propellants into the main combustion chamber (MCC) by providing a flow-field consisting of a low-velocity liquid oxygen core that is surrounded by a high-velocity, hydrogen-rich hot-gas shroud. The main injector uses 525 main injector elements and 75 baffle elements, uniformly spaced radially and circumferentially across its face.

Each main injector element assembly is made of Haynes 188 components consisting of a hollow post within a sleeve that is secured between the primary and secondary faceplates of the main injector by a retainer and nut. The posts vary in length between 6.41 and 9.01 inches, depending on their position in the injector. Each post is inertia welded to the injector body and is ported to the injector oxidizer manifold through a metering orifice in the injector body. The posts are recessed 0.23 inch from the primary faceplate to provide initial contact between the liquid oxygen stream and hot gas at a region of maximum velocity ratio. The upper end of the posts have four helically-wound spoilers machined in their outer perimeter to eliminate vibration induced by the hot-gas flow.

The retainer clamps the secondary faceplate to the sleeve and contains six equally spaced holes that direct the hot gas into the annulus surrounding the post.

The sleeve acts as a spacer between the two faceplates and, with the post, forms the annulus for the hot gas. Centering lugs are machined in the sleeve to ensure concentricity of the post and hot-gas annulus.

An A-286 CRES nut secures the primary faceplate to the sleeve and forms the cup in which the initial contact between the liquid oxygen and hot gas takes place.

304-113T

MAIN ELEMENT

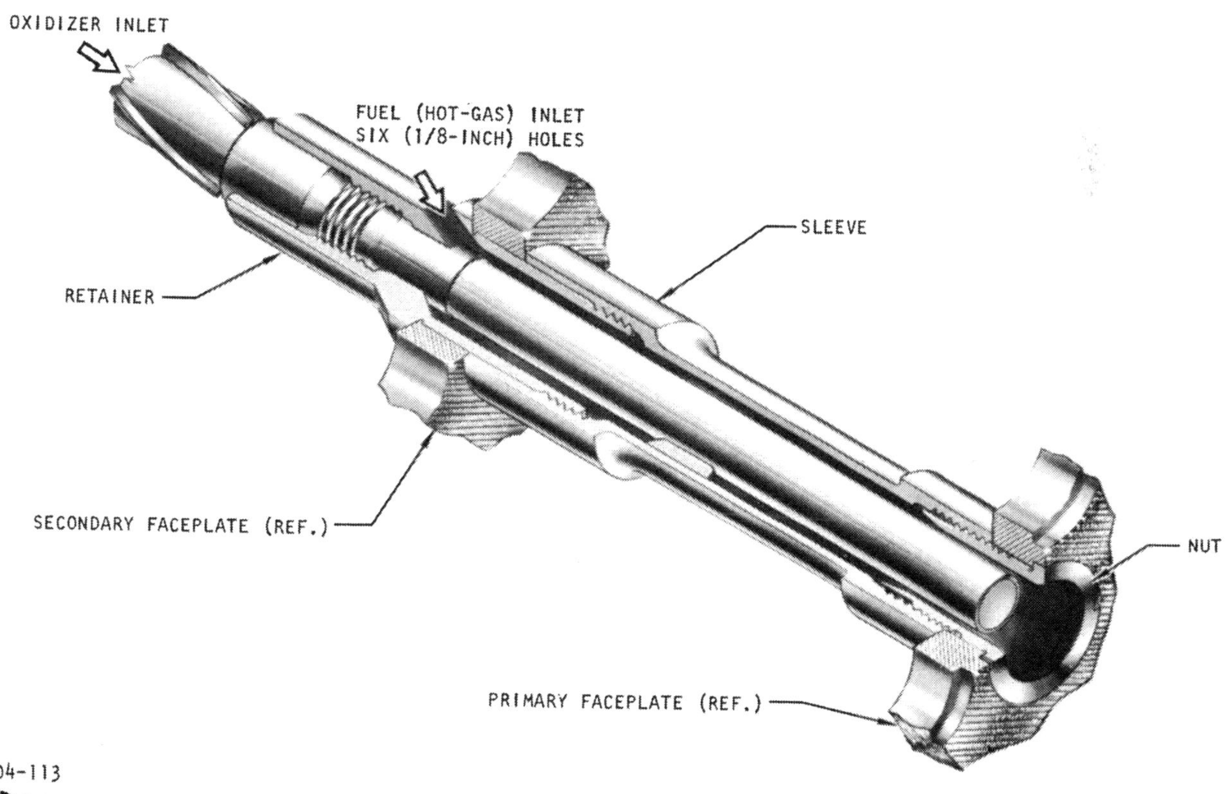

304-113

Rocketdyne Division
Rockwell International

MAIN INJECTOR BAFFLE ELEMENTS

The main injector baffle elements are bipropellant, gas/liquid, coaxial, fluid-injection devices that divide the face of the injector into six combustion compartments to prevent combustion instability modes below a frequency of 4300 Hz. Seventy-five baffle elements, along with 525 main elements, meter, mix, and distribute propellants into the MCC. The baffle elements extend 2 inches below the primary faceplate and provide a flow field consisting of a low-velocity liquid oxygen core that is surrounded by a high-velocity, cold-hydrogen gas shroud.

Each main injector baffle element assembly consists of a post and retainer (both Haynes 188), a sleeve (A-286 CRES), jacket and core (both NARloy-A). The sleeve, jacket, and core are brazed together into an integral component.

Each post is hollow and is inertia welded to the injector body and is ported to the injector oxidizer manifold through a metering orifice in the injector body. The posts vary in length between 8.95 and 11.27 inches depending on their position in the injector. The upper end of the posts have four helically-wound spoilers machined into their outer perimeter to eliminate vibration induced by the hot gas flow to the main injector elements.

The retainer threads (LH) to the posts and clamps the secondary faceplate to the sleeve which is threaded (RH) to the retainer.

The sleeve is hollow throughout its length and has 16 rows of 0.017-inch holes (512 total), which direct the cold hydrogen gas into the annulus formed by the post and sleeve. From this annulus the cold hydrogen gas is redirected through eight 0.092-inch equally spaced holes in the sleeve and into the annulus formed by the jacket and core.

The inner wall of the jacket has 32 milled rectangular channels (0.020 by 0.050 inch) to provide convective cooling of the baffle element exposed to the high temperature (6000 F) in the MCC. The geometry of the baffle tip provides both a smooth external gas transition and initiates early propellant interaction at the baffle tip.

304-114T

BAFFLE ELEMENT

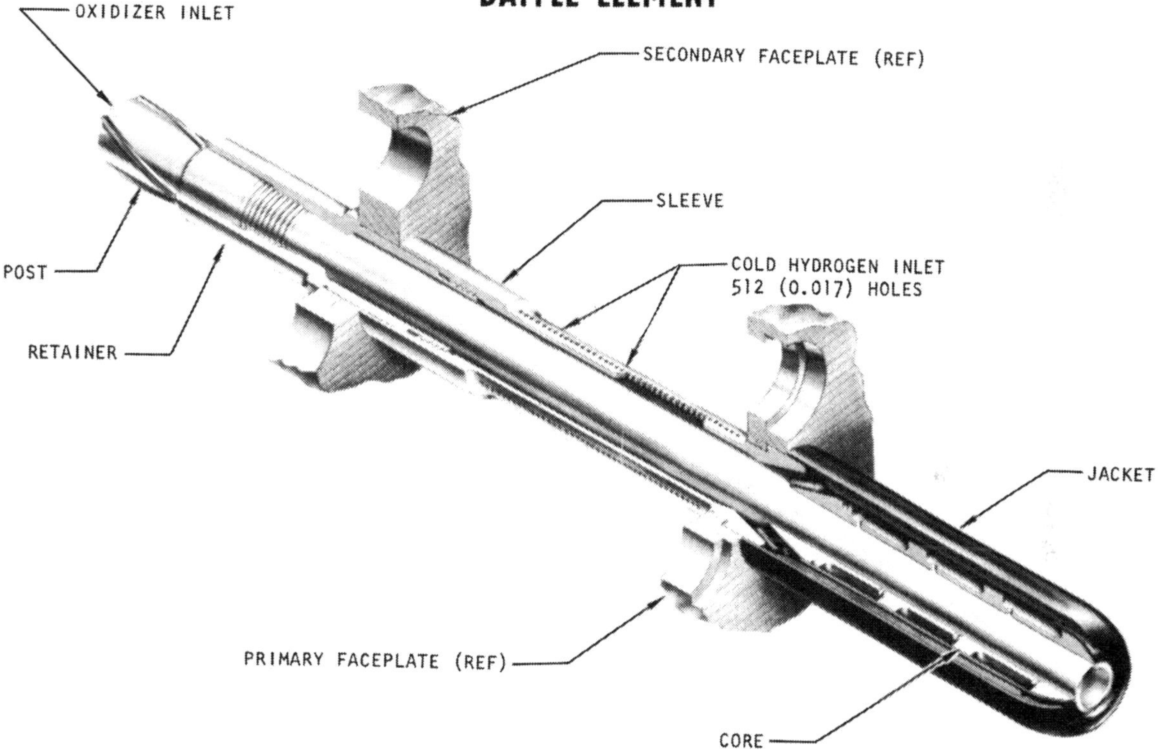

OXIDIZER INLET

SECONDARY FACEPLATE (REF)

SLEEVE

COLD HYDROGEN INLET
512 (0.017) HOLES

POST

RETAINER

JACKET

PRIMARY FACEPLATE (REF)

CORE

304-114

MAIN COMBUSTION CHAMBER

The main combustion chamber (MCC) is a cylindrical, regeneratively cooled, structural chamber that contains the burning propellant gases and initiates their expansion from the chamber throat to a ratio of 5:1. It weighs approximately 440 pounds and is flange attached to the hot-gas manifold (HGM). The MCC consists of a regeneratively fuel-cooled, NARloy-Z coolant liner and a high-strength, nickel-base alloy structural jacket.

The chamber coolant liner provides the coolant flow path for the MCC. It consists of a NARloy-Z (copper alloy) chamber machined to the same contour as the ID of the structural jacket. The outer surface has 390 milled axial coolant channels that are closed out by an electroforming process that deposits a copper barrier, followed by a nickel closeout over the coolant channels. The channels are ported to coolant inlet and outlet manifolds of the chamber jacket to provide an uppass coolant circuit for the MCC. Machined into the upper end of the liner near the injector face are 30 acoustic cavities, which damp out any high-frequency baffle compartment oscillations.

The chamber jacket provides the structural strength for the MCC. The jacket is approximately 20 inches long and is formed and machined in two matched halves. The halves are placed around the liner and welded to each other and to inlet and outlet coolant manifolds, which in turn are welded to the liner. A throat band, which provides additional structural strength, and two thrust vector control actuator strut assemblies are also welded to the jacket.

Geometry	
NARloy-Z Liner Plus Copper Barrier Plus Nickel Closeout Plus INCO 718 Structure Shell	
Number of Slots	390
Number of Acoustic Cavities	30
Injector End Diameter, inches	17.74
Throat Area, sq in.	83.41
Injector End to Throat Length, inches	14.00
Contraction Ratio	2.96:1
Expansion Ratio	5.0:1
Operating Parameters (NPL MR = 6.0)	
Injector End Pressure (Static), psia	2957
Coolant Inlet Pressure, psia	5966
Coolant Inlet Temperature, R	93
Coolant Exit Pressure, psia	4532
Coolant Exit Temperature, R	551
Coolant Flowrate, lb/sec	30.6

304-115T

MAIN COMBUSTION CHAMBER

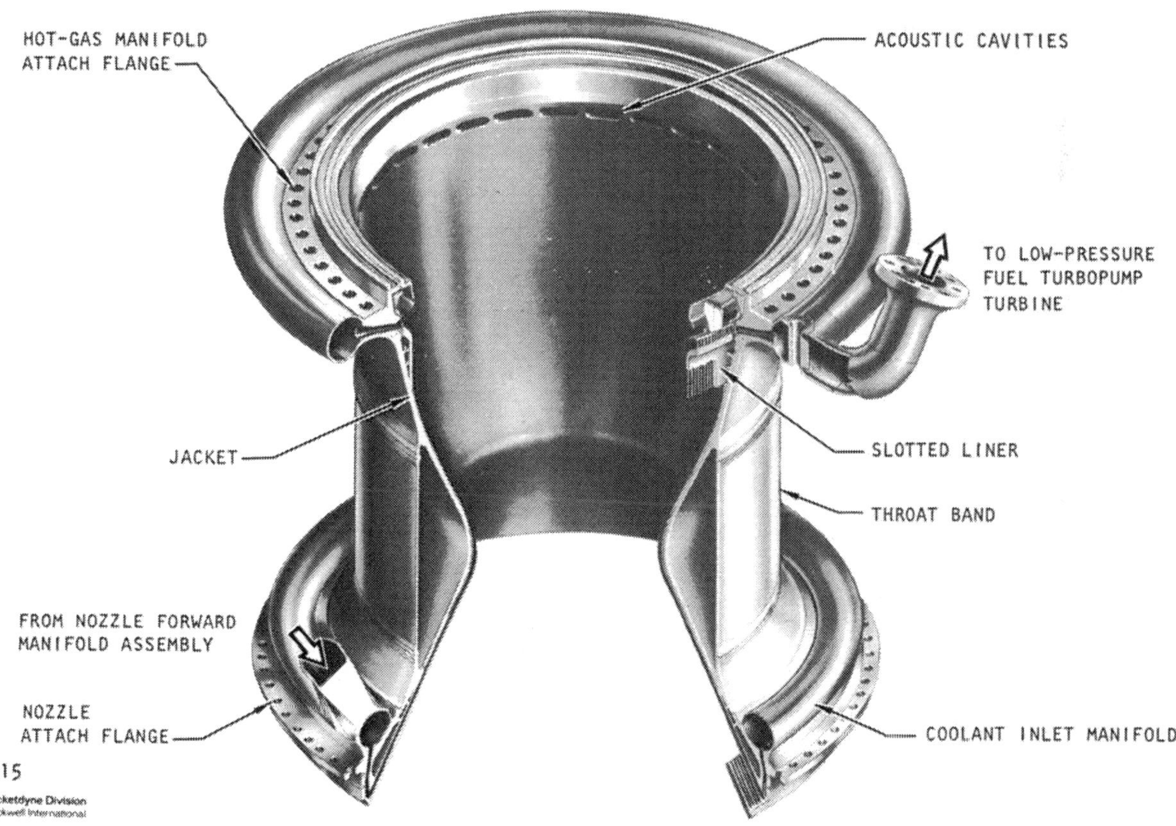

304-115

Rocketdyne Division
Rockwell International

NOZZLE ASSEMBLY

The nozzle assembly is a regeneratively fuel-cooled, 80.6-percent bell chamber that completes the expansion of the main combustion chamber (MCC) gases from a 5:1 to 77.5:1 expansion ratio. It is approximately 120 inches long, has a 94-inch exit OD, and weighs approximately 946 pounds. It is bolted to the MCC at the 5:1 expansion ratio plane.

The nozzle assembly basically consists of a forward manifold subassembly and a brazed nozzle subassembly. The forward manifold subassembly distributes hydrogen to the main chamber and nozzle cooling circuits, and provides the flanges for attaching the nozzle assembly to the MCC, main fuel valve (MFV), and preburner fuel supply duct. The nozzle assembly allows continued expansion of the combustion gases exiting from the MCC to provide the maximum possible thrust efficiency.

The forward manifold subassembly is welded to the brazed nozzle subassembly and contains one inlet and five outlets. The inlet interfaces with the downstream end of the MFV and incorporates a diffuser to efficiently distribute hydrogen to the cooling circuits when the MFV is modulated. The manifold outlets consist of three dual-outlet transfer ducts, that direct hydrogen to the nozzle coolant inlet manifold, a chamber coolant bypass duct, that contains the chamber coolant valve (CCV) to control the flowrate of hydrogen through the coolant circuits, and an MCC supply duct to direct hydrogen to the coolant inlet manifold of the MCC.

The brazed nozzle subassembly consists of 1080 A286 CRES tubes that are connected to a coolant inlet manifold at the nozzle exit and to a coolant outlet manifold at the nozzle forward end. The nozzle tubes provide an uppass cooling circuit between the inlet and outlet manifolds. The hydrogen through the circuit is directed to a mixer where it joins the hydrogen flow through the CCV and is used as a fuel by the preburners. The forward section of the nozzle is enclosed in an INCO 718 jacket and insulated hat bands are spaced and welded over the length of the nozzle to provide the required hoop strength.

Geometry	
Attach Point Area Ratio	5.0:1
Exit Area Ratio	77.5:1
Length (Throat To Exit), Inches	121
Exit Diameter (Inside/Outside), Inches	90.7/94
Number Of Tubes	1080
Number Of Feed Ducts	3
Operating Parameters (NPL MR = 6.0)	
Coolant Flowrate, lb/sec	38.3
Coolant Inlet	
Temperature, R	93
Pressure, psia	5993
Coolant Exit	
Temperature, R	684
Pressure, psia	5847

NOZZLE ASSEMBLY

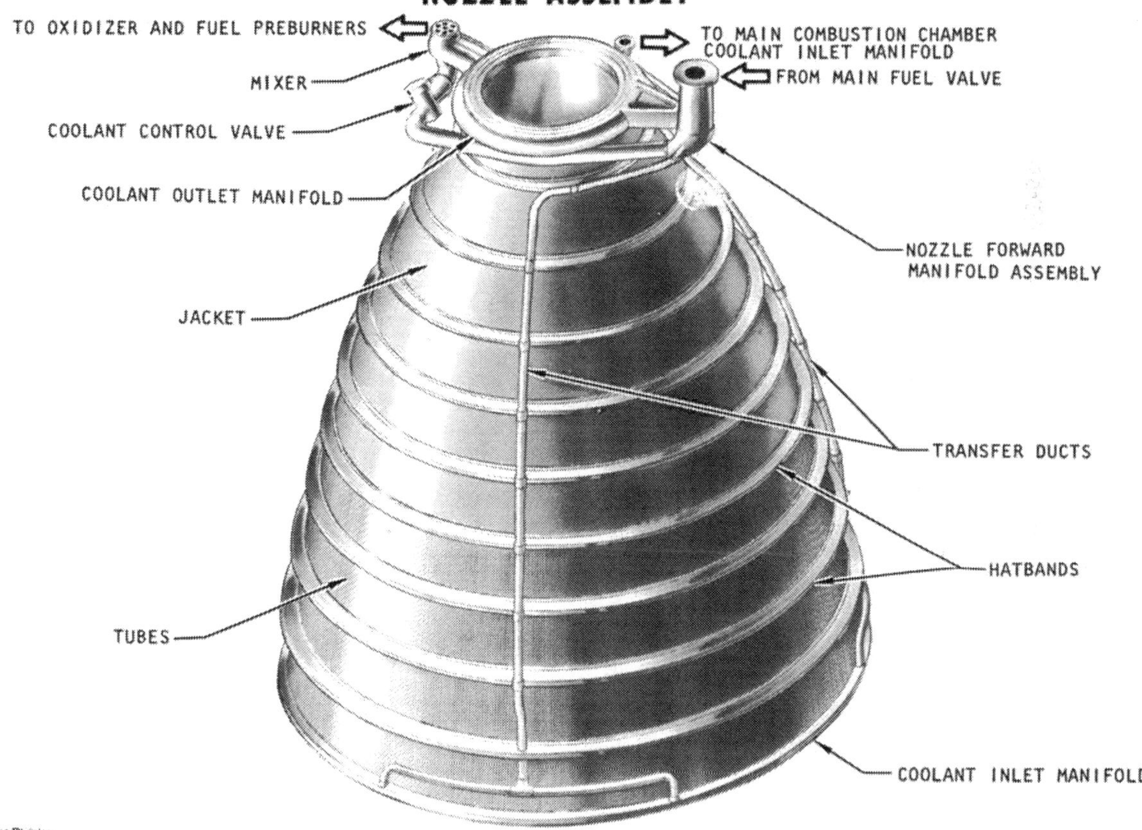

304-116

Rocketdyne Division
Rockwell International

LOW-PRESSURE OXIDIZER TURBOPUMP

The low-pressure oxidizer turbopump (LPOTP) is an axial-flow pump driven by a six-stage turbine that is powered by liquid oxygen. During engine start and mainstage, the LPOTP maintains sufficient pressure to the high-pressure oxidizer turbopump (HPOTP) to permit the HPOTP to operate at high speeds without an inducer and without cavitation even in the worst case of engine inlet conditions. Turbine-drive fluid is tapped from the HPOTP discharge and, after powering the turbine, is injected into the pumped fluid through a port between the turbine discharge and pump discharge volutes. The combined flows are then routed to the HPOTP inlet. Since the pumped fluid and turbine drive fluid are both liquid oxygen, the requirement for dynamic seals, purges, and drains has been eliminated.

The rotor is supported by two liquid oxygen-cooled ball bearings. The turbine end bearing coolant flow path is from the last stage of the turbine, through the bearing, hollow rotor, radial holes in the rotor, and to the inducer discharge. Coolant for the inducer-end bearing is from the turbine inlet, through the rotor labyrinth seal, through the bearing, and to the inducer discharge.

A redundant-element, magnetic-type speed transducer is installed on the turbine end of the turbopump housing.

Access ports in the vehicle duct allow periodic inspection of the inducer blades for damage. Inspection is performed with a fiber optic borescope.

Major parts of the turbopump are a TENS-50 aluminum alloy housing; K-Monel nickel-alloy inducer, housing inlet liner, rotor, and stator assembly; 440C CRES ball bearings with Teflon-impregnated fiberglass cages; and a silver rotor labyrinth seal. The LPOTP has an approximate overall dimensional envelope of 18 by 18 inches and weighs approximately 185 pounds.

NPL (MR = 6.0)

Parameter	Value	Parameter	Value
Inlet Flowrate, lb/sec	883.9	Turbine Flowrate, lb/sec	164.1
Discharge Flowrate, lb/sec	1047.9	Turbine Inlet Pressure, psia	4481.0
Discharge Pressure, psia	415.1	Turbine Discharge Pressure, psia	414.9
Discharge Temperature, R	168.9	Turbine Inlet Temperature, R	187.9
Brake Horsepower	1468.0	Turbine Discharge Temperature, R	187.0
Speed, rpm	5146.0		

304-117T

LOW PRESSURE OXYGEN TURBOPUMP

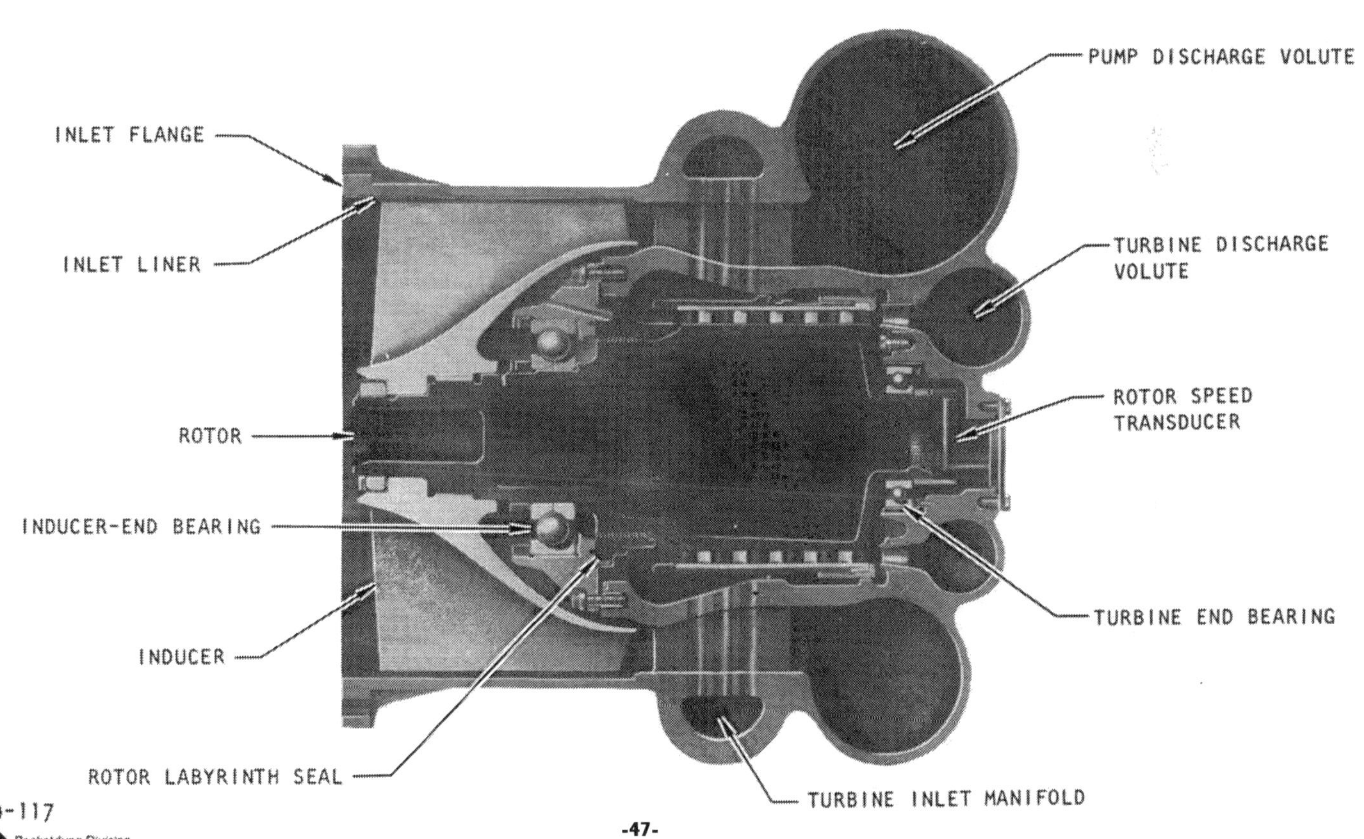

304-117

Rocketdyne Division
Rockwell International

LOW-PRESSURE FUEL TURBOPUMP

The low-pressure fuel turbopump (LPFTP) is an axial-flow pump driven by a two-stage turbine that uses gaseous hydrogen as the power medium. During engine start and mainstage operation, the LPFTP maintain sufficient pressure to the high-pressure fuel turbopump (HPFTP) to permit the HPFTP to operate at high speeds without an inducer and without cavitation even in the worst case of engine inlet conditions. The turbine is driven by gaseous hydrogen from the main combustion chamber (MCC) coolant outlet manifold.

The inducer and shaft are supported by three liquid hydrogen-cooled ball bearings. The bearing coolant is tapped from the pump volute and flows in series through the turbine-end and inducer-end bearings. The coolant is then returned to the pump inlet through holes in the inducer hub.

Three shaft seals are used to control leakage between the pump and turbine before engine start and during operation. Before engine start, leakage from the pump into the turbine is prevented by a spring-loaded-closed, propellant pressure-actuated-open, lift-off seal. During engine start the seal nose is separated from its mate ring when increasing fuel pressure overcomes the spring force. A positive separation between the seal nose and mate ring is maintained until engine shutdown when fuel pressure decreases below spring force. During operation, leakage from the turbine into the pump is minimized by the pump and turbine seals. These seals are controlled-gap, floating-ring-type dynamic seals. The pump seal is an Inconel band with a carbon insert that interfaces with a chrome-plated area on the shaft. The turbine seal is an Inconel band with an Amcermet 701-65 (a sintered Inconel matrix that is infiltrated with barium fluoride-calcium fluoride as a lubricant) insert that interfaces with a chrome-plated area on the shaft.

A redundant-element, magnetic-type speed transducer is installed in the pump volute to monitor shaft speed.

Access ports in the vehicle duct allow periodic inspection of the inducer blades for damage. Inspection is performed with a fiber optic borescope.

Major parts of the LPFTP are: The TENS-50 aluminum alloy pump housing; nickel-base alloy turbine manifold; titanium alloy inducer; A-286 CRES shaft, turbine disk, turbine blades, and stators; and 440C CRES ball bearings with Armalon cages. The pump housing is insulated with polyurethane foam that is protected by an electroformed shell.

The LPFTP has an overall dimensional envelope of approximately 18 by 24 inches and weighs approximately 130 pounds.

NPL (MR = 6.0)

Parameter	Value	Parameter	Value
Inlet Flowrate, lb/sec	147.3	Turbine Flowrate, lb/sec	30.6
Discharge Flowrate, lb/sec	147.3	Turbine Inlet Pressure, psia	4257.0
Discharge Pressure, psia	233.5	Turbine Discharge Pressure, psia	3589.0
Discharge Temperature, R	39.4	Turbine Inlet Temperature, R	550.7
Brake Horsepower	2400.0	Turbine Discharge Temperature, R	539.0
Speed, rpm	14,777.0		

LOW PRESSURE FUEL TURBOPUMP

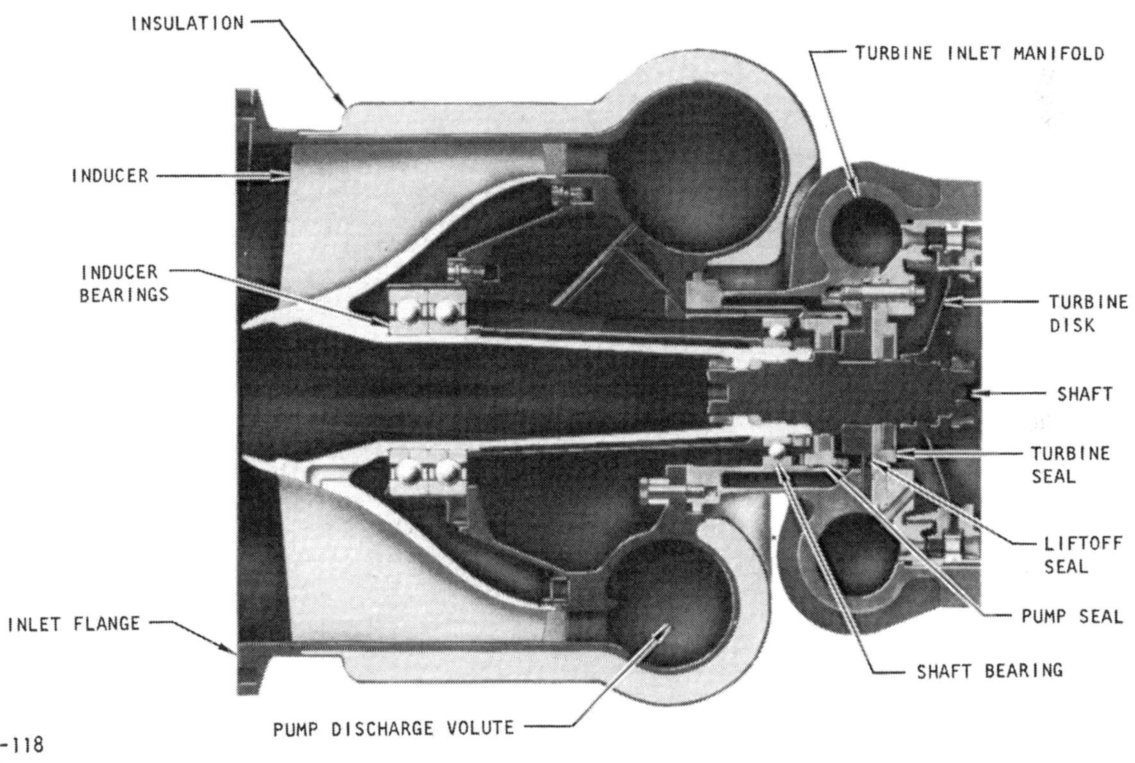

304-118

Rocketdyne Division
Rockwell International

HIGH-PRESSURE OXIDIZER TURBOPUMP

The high-pressure oxidizer turbopump (HPOTP) consists of two single-stage centrifugal pumps on a common shaft that are directly driven by a two-stage hot-gas turbine. The main pump receives oxidizer from the low-pressure oxidizer turbopump (LPOTP) and supplies oxygen at an increased pressure to the LPOTP turbine, the heat exchanger, the preburner pump, and the thrust chamber injector. The preburner pump further increases the pressure to the level required by the oxidizer preburner (OPB) and fuel preburner (FBP). The turbine is powered by hot gas (hydrogen-rich steam) supplied by the OPB. Mixing of oxidizer and hot gas is prevented by seals, a purge, and drains. Two duplex sets of liquid oxygen-cooled ball bearings support the rotating parts. The HPOTP is flange attached to the hot-gas manifold (HGM) and is canted at a 10-degree angle out from the engine centerline. The HPOTP is a line replaceable unit (LRU) with an overall dimensional envelope of approximately 24 by 36 inches. It weighs approximately 530 pounds.

NPL (MR = 6.0)

Main Pump Flowrate, lb/sec	1047.8
Main Pump Inlet Temperature, R	168.9
Main Pump Discharge Temperature, R	187.9
Main Pump Inlet Pressure, psia	359.8
Main Pump Discharge Pressure, psia	4624.2
Main Pump Brake Horsepower	21,297
Speed, rpm	29,225
Preburner Pump Flowrate, lb/sec	104.4
Preburner Inlet Temperature, R	187.9
Preburner Pump Discharge Temperature, R	205.9
Preburner Pump Inlet Pressure, psia	4426.5
Preburner Pump Discharge Pressure, psia	7634.0
Preburner Pump Brake Horsepower	1490
Turbine Flowrate, lb/sec	56.3
Turbine Inlet Pressure, psia	5163.0
Turbine Discharge Pressure, psia	3358.0
Turbine Inlet Temperature, R	1405.0
Turbine Discharge Temperature, R	1283.0
Turbine Brake Horsepower	22,787

304-119T

HIGH PRESSURE OXYGEN TURBOPUMP

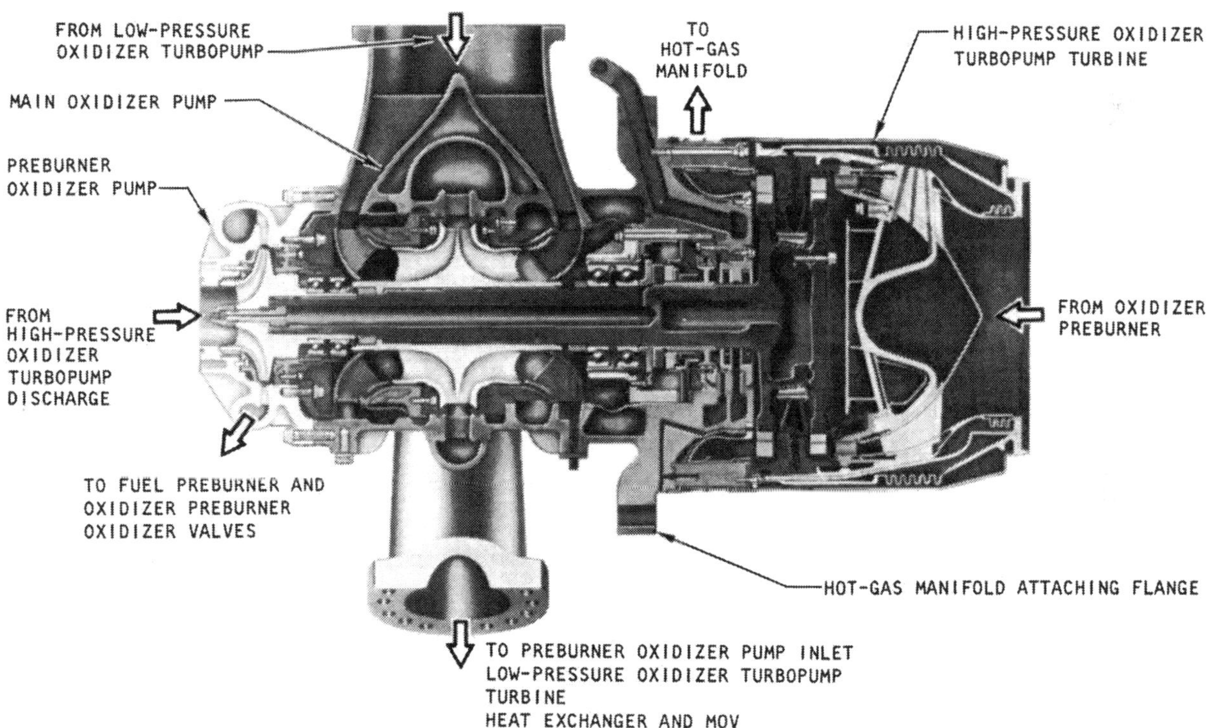

304-119

HIGH-PRESSURE OXIDIZER MAIN PUMP

The high-pressure oxidizer main pump has a single inlet with a 50-50 flow split into a double-entry common outlet impeller. Liquid oxygen enters the main pump through the main pump housing where the flow split is made. Inlet vanes direct the flow to the impeller inlet guide vanes, which in turn direct the flow to the impeller inlets. The impeller has four full and four partial blades in each half. After passing through the impeller, the flow is redirected into the discharge volute by diffuser vanes.

Turbopump shaft bearings are cooled by liquid oxygen from the preburner pump. Coolant for the preburner-end bearings is through the preburner pump impeller hub labyrinth seal, through the bearings, and to the main pump impeller inlet. The turbine-end bearings coolant is through the preburner impeller bolt, through the hollow shaft, through the bearings, and to the main pump impeller inlet. Pump shaft axial thrust is balanced in that the double-entry main impeller is inherently balanced and the thrusts of the preburner pump and turbine are equal but opposite. Residual shaft thrust is controlled by a self-compensating, nonrubbing, balance piston function by using the main impeller seal leakage flows and controlling the pressure on the impeller shrouds by orifices at the impeller discharge tips. Mixing of oxidizer and turbine gas is prevented by a dynamic shaft seal package that is between the main pump and the turbine. The seal package consists of an oxidizer labyrinth seal, a hydrodynamic primary oxidizer seal, a pressure-actuated controlled-gap intermediate seal, and two controlled-gap turbine hot-gas seals. Drain cavities with overboard drain lines are located between the primary oxidizer seal and the intermediate seal, between the intermediate seal and secondary turbine seal, and between the secondary and primary turbine seals. To further ensure against the mixing of oxidizer and turbine gas, a helium purge is applied between the elements of the intermediate seal during engine operation.

The preburner pump has a single-entry impeller that discharges oxidizer through diffuser vanes into the discharge volute. The preburner pump housing is flange mounted to the main pump housing. Impeller seals interface with labyrinths cut in the outside diameter of the impeller shroud and hub to minimize leakage back to the pump inlet and control the flow for cooling the preburner pump end bearings.

304-120T

HIGH-PRESSURE OXYGEN TURBOPUMP

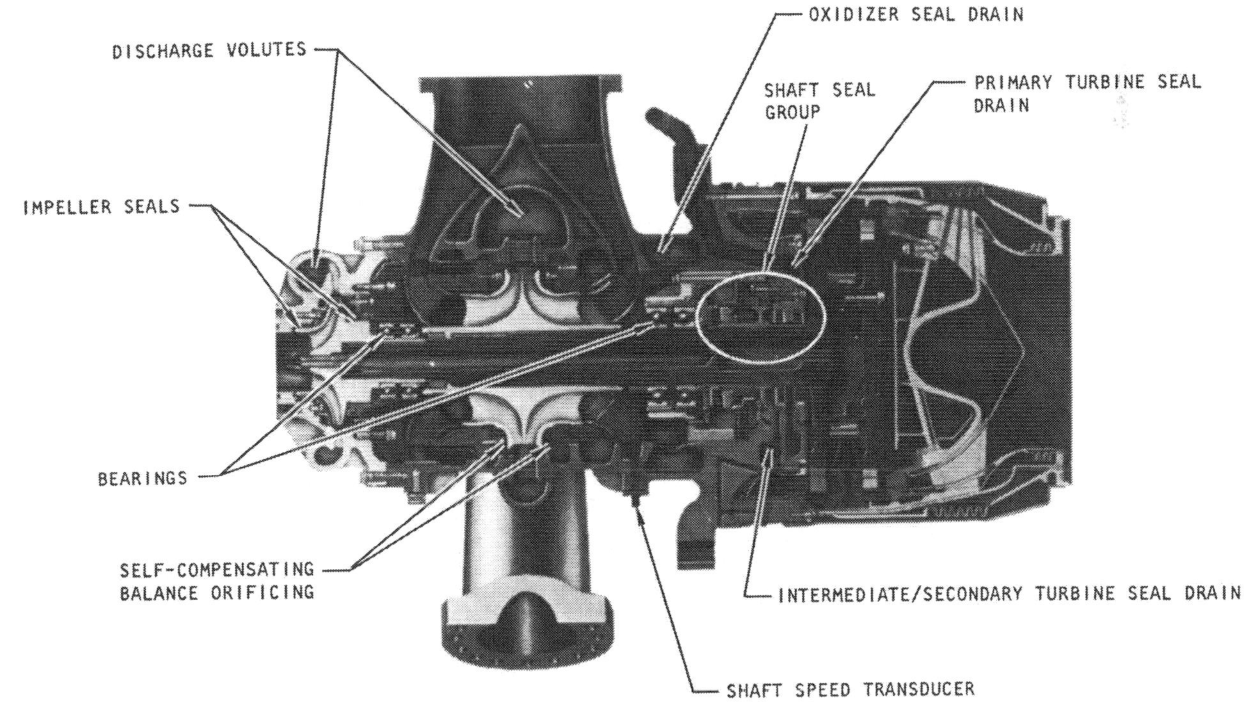

OXIDIZER SEAL DRAIN

DISCHARGE VOLUTES

SHAFT SEAL GROUP

PRIMARY TURBINE SEAL DRAIN

IMPELLER SEALS

BEARINGS

SELF-COMPENSATING BALANCE ORIFICING

INTERMEDIATE/SECONDARY TURBINE SEAL DRAIN

SHAFT SPEED TRANSDUCER

304-120

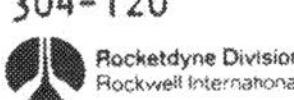

HIGH-PRESSURE OXIDIZER TURBOPUMP TURBINE

The high-pressure oxidizer turbopump turbine is powered by hot gas generated by the oxidizer preburner (OPB). Hot gas enters the turbine and flows across the shielded support struts, through the first-, and second-stage nozzles and blades, and is discharged into the hot-gas manifold (HGM). The turbine wheels are mated through a curvic coupling and are held together with a circle of bolts. The second-stage wheel is integral with the pump shaft. Turbine blade-to-housing leakage is minimized by lands on the outer perimeter of the blade shrouds that run against seals in the turbine housing.

All components of the turbine are cooled by gaseous hydrogen flowing over or through them. Coolant is supplied from the OPB coolant jacket. After cooling the turbine components, the coolant is exhausted into the hot-gas flow stream.

Turbine-to-OPB sealing is accomplished by a pair of concentric bellows that load dual seals in the turbine inlet flange to the OPB. Turbine component materials are primarily of nickel- and cobalt-base alloys. The diameter of the turbine wheels with blades is approximately 11 inches.

304-121T

HIGH PRESSURE OXYGEN TURBOPUMP

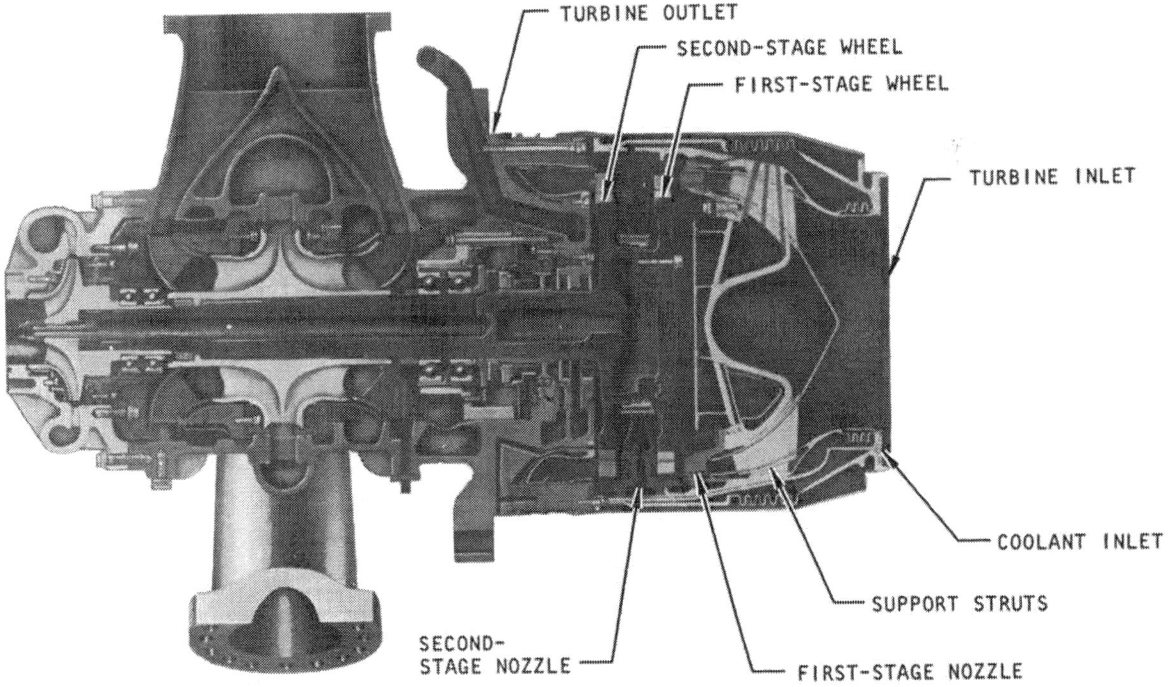

TURBINE OUTLET

SECOND-STAGE WHEEL

FIRST-STAGE WHEEL

TURBINE INLET

COOLANT INLET

SUPPORT STRUTS

SECOND-
STAGE NOZZLE

FIRST-STAGE NOZZLE

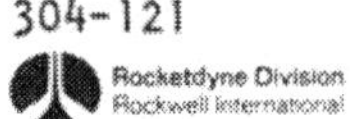

HIGH-PRESSURE FUEL TURBOPUMP

The high-pressure fuel turbopump (HPFTP) is a three-stage centrifugal pump that is directly driven by a two-stage hot-gas turbine. The pump receives fuel from the low-pressure fuel turbopump (LPFTP) and supplies it at increased pressure, through the main fuel valve (MFV), to the thrust chamber assembly coolant circuits. The turbine is powered by hot gas (hydrogen-rich steam) generated by the FPB.

Fuel leakage into the turbine before engine start is prevented by a pressure-actuated static lift-off shaft seal. During engine operation, leakage is controlled by dynamic seals. Two duplex sets of liquid hydrogen-cooled ball bearings support the rotating parts. The HPFTP is flange attached to the hot-gas manifold (HGM) and is canted out from the engine centerline at a 10-degree angle. The HPFTP is a line replaceable unit (LRU) with an overall dimensional envelope of approximately 22 by 44 inches. It weighs approximately 705 pounds.

NPL (MR = 6.0)

Parameter	Value	Parameter	Value
Pump Flowrate, lb/sec	147.3	Turbine Inlet Pressure, psia	5161
Pump Inlet Pressure, psia	177.5	Turbine Discharge Pressure, psia	3381
Pump Discharge Pressure, psia	6189.2	Turbine Inlet Temperature, R	1729
Pump Inlet Temperature, R	39.4	Turbine Discharge Temperature, R	1584
Pump Discharge Temperature, R	93.1	Brake Horsepower	62,241
Turbine Flowrate, lb/sec	142.7	Speed, rpm	35,082

304-122T

HIGH-PRESSURE FUEL TURBOPUMP

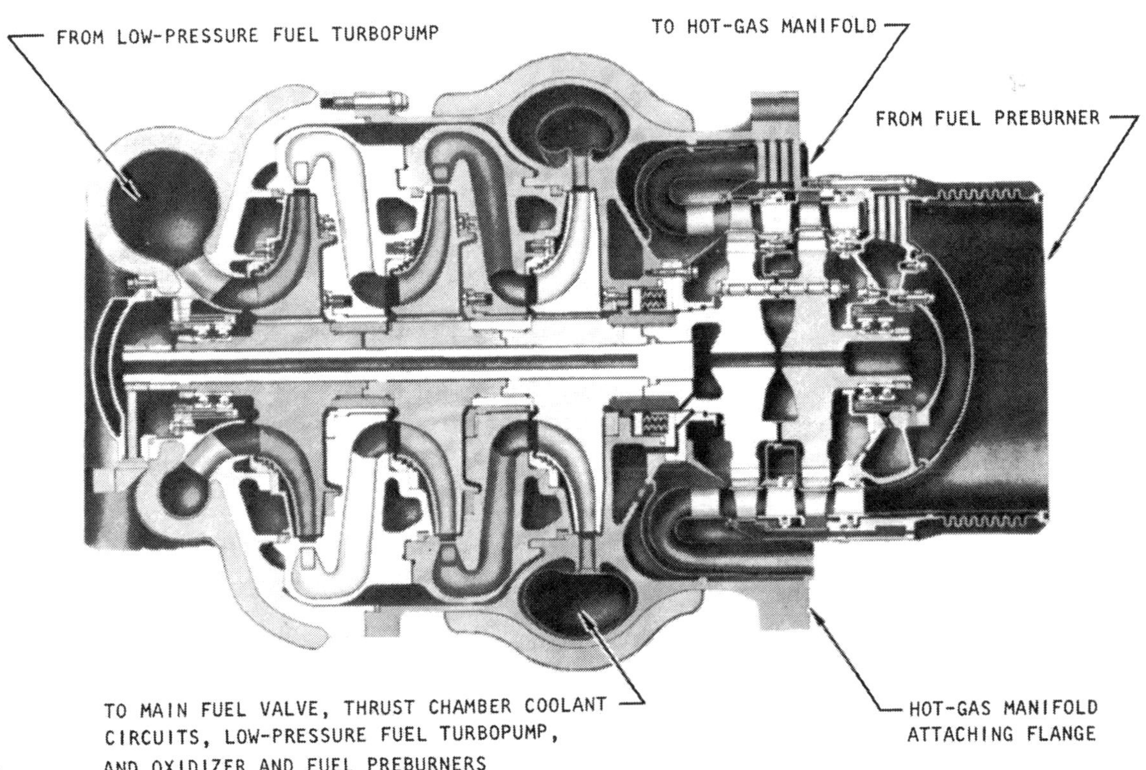

304-122

HIGH-PRESSURE FUEL PUMP

The high-pressure fuel pump is a three-stage pump with a titanium alloy inlet housing, a nickel-base alloy main housing, three titanium alloy impellers, two aluminum alloy interstage diffusers, three nickel-base alloy splined sleeves, two sets of fuel-cooled ball bearings, a static shaft seal, and eight labyrinth-type seals. The pump is insulated with polyurethane foam that is protected by electroformed nickel shells. Fuel flows in series through the three impellers from the pump inlet to outlet with flow being redirected between stages by interstage diffusers. The three impellers have 6 full blades, 6 long partial blades, and 12 short partial blades.

Coolant for the shaft pump-end bearing is tapped off between the pump first and second stages and ported between the impeller ID and the tie bolt. Coolant flows through the bearings and returns to the inlet of the first-stage impeller. Coolant flow to the shaft turbine-end bearings is supplied from the pump balance piston cavity through the shaft static lift-off seal, and to the bearings.

Axial rotor thrust is controlled by a self-compensating, double-acting balance piston that operates between high-pressure and low-pressure orifices to maintain the thrust at zero.

Before engine start, leakage from the pump into the turbine is prevented by a spring-loaded closed/propellant pressure-actuated open, lift-off seal. During engine start the seal nose is separated from its mate ring when increasing fuel pressure overcomes the spring force. A positive separation between the seal nose and mate ring is maintained until engine shutdown when fuel pressure decreases below spring force. Propellant flow through the seal and mate ring is used to cool the turbine-end bearings and the turbine components.

304-123T

HIGH-PRESSURE FUEL TURBOPUMP

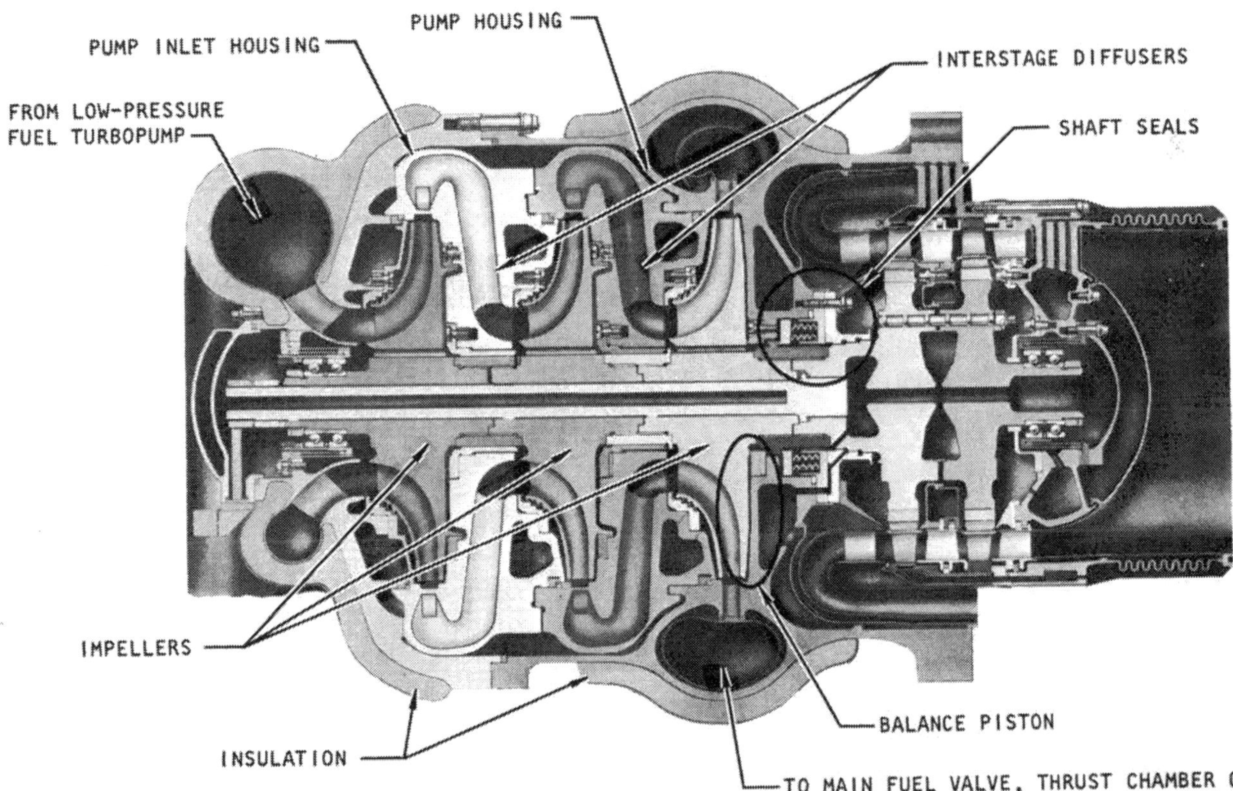

304-123

Rocketdyne Division
Rockwell International

HIGH-PRESSURE FUEL TURBOPUMP TURBINE

The high-pressure fuel turbopump turbine is powered by hot-gas generated by the fuel preburner (FPB). Hot gas enters the turbine and flows across the shielded support struts, through the first- and second-stage nozzles and blades, and is discharged into the hot-gas manifold (HGM). The turbine wheels are mated through a curvic coupling and are held together with a circle of bolts. The two-stage turbine transmits torque to the pump by a splined coupling between the second-stage wheel and the pump third-stage impeller.

Bearing and turbine coolant is supplied through the shaft static lift-off seal when the seal is actuated at engine start. The turbine coolant flows over or through all hot-gas components and is then exhausted into the hot-gas flow stream.

Turbine-to-FPB sealing is accomplished by a bellows that loads a Naflex seal against the FPB flange.

Turbine components materials are primarily of nickel- and cobalt-base alloys.

304-124T

HIGH-PRESSURE FUEL TURBOPUMP

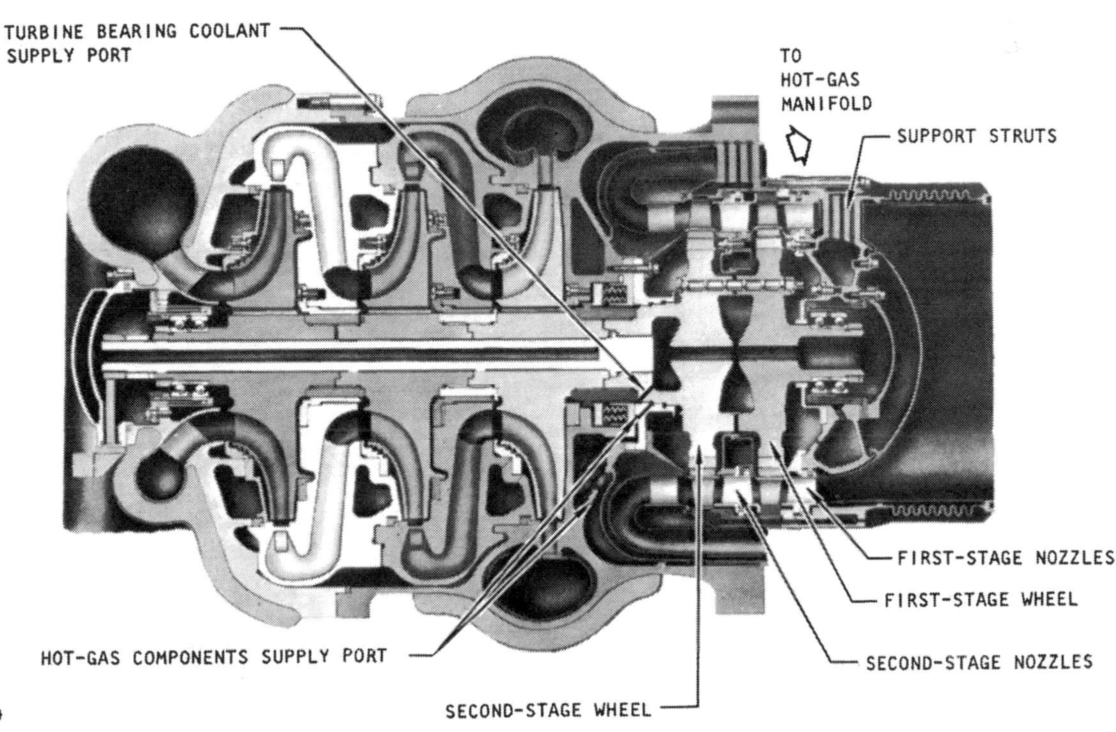

304-124

Rocketdyne Division
Rockwell International

INTERCONNECTS

Engine interconnects are divided into three categories: main propellant articulating ducts, fluid interface lines, and component interconnects.

Main propellant articulating ducts interconnect the nongimbaled low-pressure turbopumps to components of the engine that gimbal. The ducts are of the wraparound type and have four rigid sections and three flexible sections. The flexible sections consist of a convoluted bellows, a bellows sleeve, and a tension restraint. Three types of tension restraints are used. One has external yokes and gimbal ring with shear pins, one has internal tripods with a mating ball and socket, and one has an internal ball and socket.

Fluid interface lines are the vehicle-to-engine lines for recirculation of propellants, propellant tank pressurants, hydraulics, and pneumatics. Three types of lines are used: articulating flexible lines, flexible hoses with overbraided convoluted bellows sections, and a rigid line.

Component interconnects are rigid lines with the exception of one small-diameter flexible hose.

Where required, fuel ducts and lines are vacuum-jacketed to prevent liquid air condensation and help to maintain the fuel temperature at the desired level.

All interconnects with separable connections have bolted flange joints. The basic flange configuration is a modified ASME design with a pilot adjacent to the fluid passage. The pilot design maintains a direct in-line load path at the flange ID under all operating conditions to provide positive clamping of the joint seal.

304-125T

INTERCONNECTS

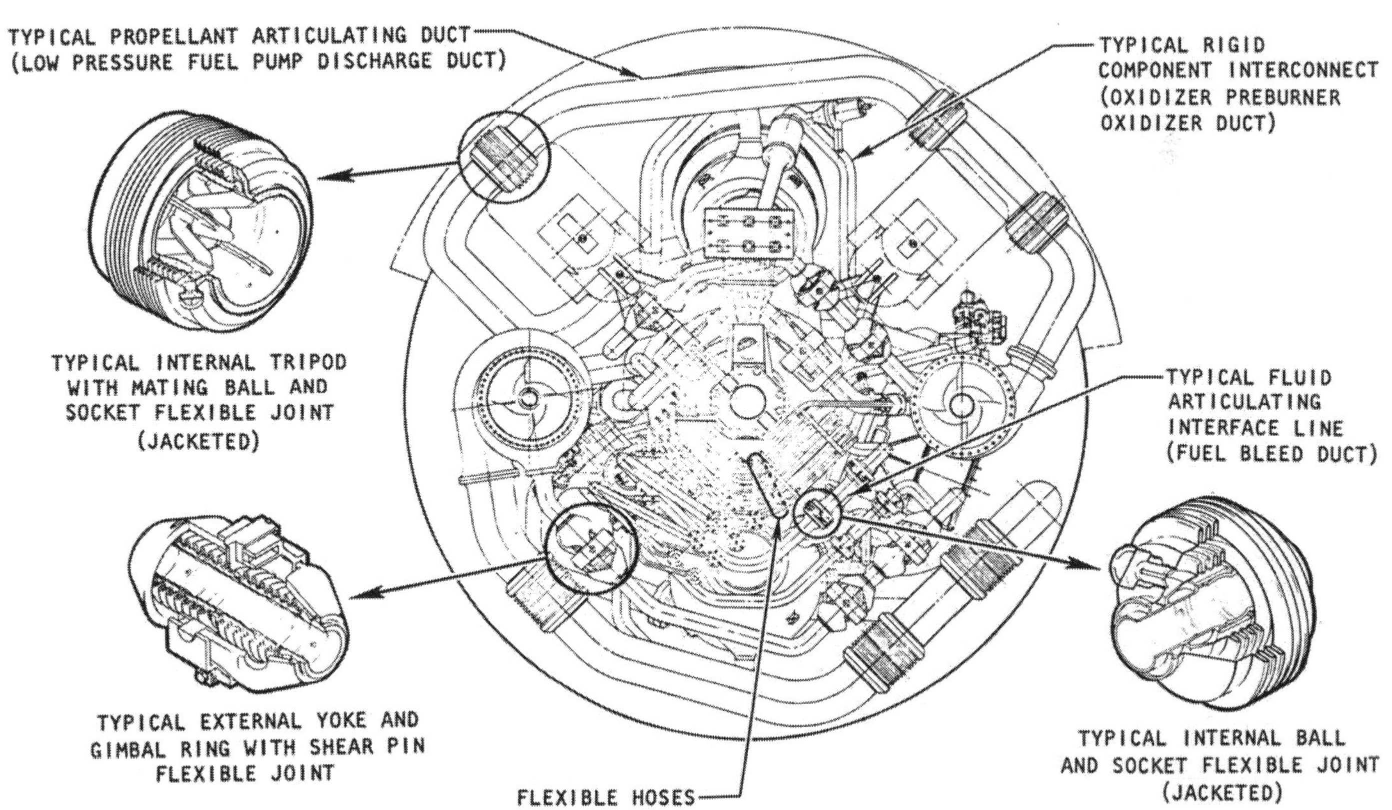

304-125

HEAT EXCHANGER

The heat exchanger is a multipath, single-pass, coilpack installed in the oxidizer side of the hot-gas manifold (HGM). It converts liquid oxygen to gaseous oxygen for vehicle oxygen tank pressurization. The heat exchanger weighs approximately 20 pounds and consists of a helically wound, small tube (0.1875-inch ID) approximately 30.6 inches long, in series with two parallel, larger tubes (0.375-inch ID) each approximately 310 inches long. The tubes are Rene 41 material and are attached to supports welded to the HGM coolant jacket inner wall. The cross-flow of hot turbine exhaust gases from the high-pressure oxidizer turbopump (HPOTP) heats the liquid oxygen to a gas.

Liquid oxygen, tapped off the discharge side of the HPOTP, is supplied to the inlet of the heat exchanger through an antiflood valve. The oxygen is heated to a gas in the small tube (first stage) and to the final outlet temperature in the two larger tubes (second stage). An orificed bypass line around the heat exchanger injects an unheated portion (approximately 30 percent) of the total oxygen flow into the outlet of the heat exchanger for control of temperature/flowrate operating characteristics. Orifices in the vehicle control heat exchanger flowrate.

At NPL (MR = 6.0), the heat exchanger delivers 2.25 lb/sec (including 0.65 lb/sec bypass flow) at a temperature and pressure of 390 F and 3238 psia.

	Oxygen	Hot Gas
Operating Conditions (NPL MR = 6.0)		
Flowrate, lb/sec	2.25	57.6
Temperature, R		
Inlet	188	1243
Outlet	850	1241
Pressure, psia		
Inlet	4380	3348
Outlet	3260	3307

304-126T

LOX TANK PRESSURIZATION HEAT EXCHANGER

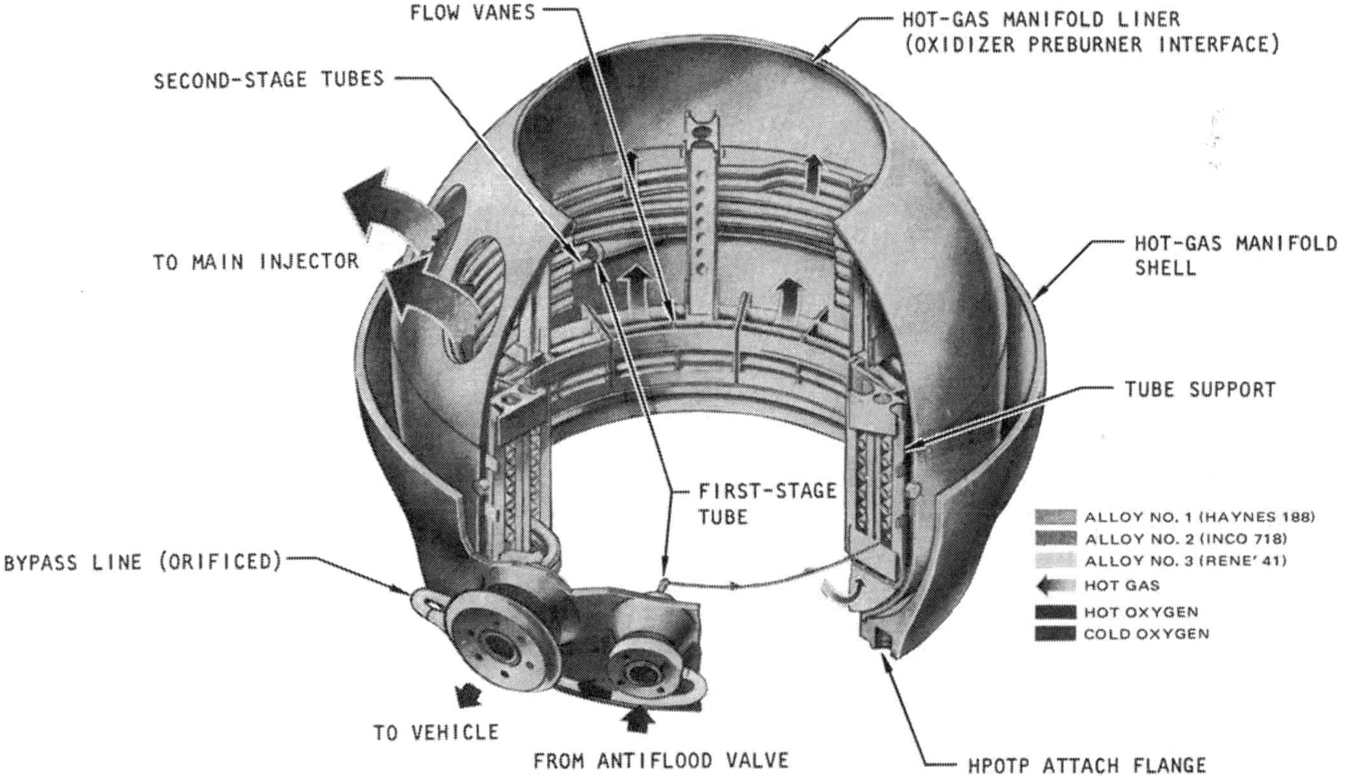

FLOW VANES

HOT-GAS MANIFOLD LINER
(OXIDIZER PREBURNER INTERFACE)

SECOND-STAGE TUBES

TO MAIN INJECTOR

HOT-GAS MANIFOLD
SHELL

TUBE SUPPORT

FIRST-STAGE
TUBE

ALLOY NO. 1 (HAYNES 188)
ALLOY NO. 2 (INCO 718)
ALLOY NO. 3 (RENE' 41)
HOT GAS
HOT OXYGEN
COLD OXYGEN

BYPASS LINE (ORIFICED)

TO VEHICLE

FROM ANTIFLOOD VALVE

HPOTP ATTACH FLANGE

304-126

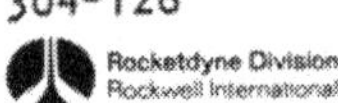

Rocketdyne Division
Rockwell International

The main oxidizer valve (MOV) is a ball-type valve with a 2.5-inch propellant flow passage flange-mounted between the main chamber oxidizer dome and the high-pressure oxidizer duct. The valve controls oxidizer flow to the main chamber LOX dome and main chamber augmented spark igniter (ASI), and is operated by a hydraulic servoactuator mounted to the valve housing. The servoactuator receives its control signals from the controller.

Basically, the valve consists of three moving components: a ball, a shaft, and a ball inlet seal. The ball, the shaft, and a preloaded torsional spring are interlocked by pins so that a spring closing force is applied to the ball. The ball inlet seal is a machined Kel-F, bellows-loaded seal that prevents leakage around the ball when the valve is closed. Redundant shaft seals prevent leakage along the shaft (actuator end) during engine operation. Inlet and outlet throttling sleeves align flow passages in the ball and shaft to minimize flow turbulence and pressure losses. Erosion of the inlet seal is prevented by keeping the inlet sleeve edge close to the ball. Five low-friction roller bearings reduce the force required to actuate the valve. Two are located between the ball and shaft, two between the cams and the valve housing, and a thrust bearing between the shaft seal and upper bearing.

The MOV is initially ramped open at engine start to establish ignition oxidizer flow and a low engine power level. As engine thrust builds up, the valve is ramped to the full open position at a rate that achieves a smooth transient operation of the engine within the required maximum thrust increase rate. The valve is also ramped closed. These ramp rates ensure that the engine starts and shuts down without thrust overshoot or undershoot. During the first 15 degrees of valve opening travel, two cams rotate to actuate a cam follower mechanism that retracts the inlet seal. This feature eliminates rubbing between the seal and ball and prolongs seal life. As the seal is retracted, oxidizer flows to the main chamber dome and ASI to establish ignition flow. After this initial shaft movement, the ball is opened by positive engagement with the shaft to establish oxidizer flow for power buildup. On closing, the ball moves with the shaft until the ball engages a mechanical stop. The shaft continues to rotate an additional 15 degrees, during which time the cam allows the seal to reseat on the ball.

The ball, shaft, housing, cam, spring, bellows, link pins, and shaft end nuts are made of INCO 718, which provides outstanding cryogenic properties, is weldable, and is not susceptible to stress corrosion.

The valve bearings and seal cam rollers are 440C CRES, which provides high strength, stability, and wear resistance for these precision parts. The bearing race retainers and shaft end locking washers are 302 CRES.

The MOV weighs approximately 100 pounds.

Geometry	
Inconel 718 Housing, Shaft, Ball and Bellows 440C Shaft Seal Seat	
Flow Passage Diameter, Inches	2.500
Ball Diameter, Inches	5.125
Shaft Diameter (Seal), Inches	1.875
Valve Length, Inches	10.25
Operating Parameters (NPL MR = 6.0)	
Inlet Pressure, psia	4409
Outlet Pressure, psia	4368
Temperature, R	188
Flowrate, lb/sec	794.8

SSME MAIN OXIDIZER VALVE ASSEMBLY

(TYPICAL FOR MAIN FUEL VALVE, FUEL PREBURNER OXIDIZER VALVE, AND OXIDIZER PREBURNER OXIDIZER VALVE)

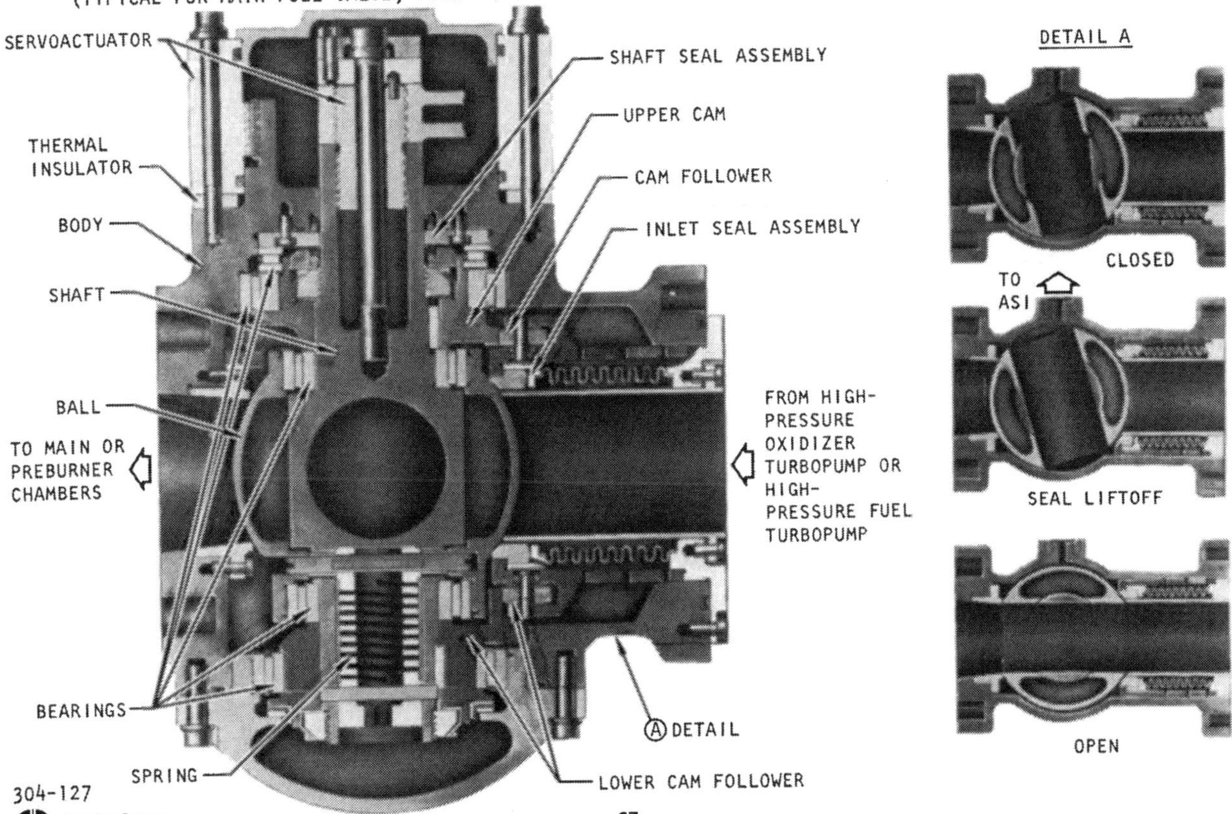

304-127

MAIN FUEL VALVE

The main fuel valve (MFV) is a hydraulically actuated, spring-loaded-closed, ball-type valve with a 2.5-inch propellant flow passage. The MFV is flange-mounted between the high-pressure fuel duct and coolant inlet distribution manifold of the thrust chamber nozzle. The valve permits or stops the flow of fuel to the thrust chamber augmented spark igniter (ASI), thrust chamber coolant circuits, the low-pressure fuel turbopump (LPFTP) turbine, hot-gas manifold (HGM) coolant circuit, and the oxidizer preburner (OPB), and fuel preburner (FPB). Valve position is controlled by commands from the engine controller.

Major parts of the valve are a housing, shaft, ball, bellows assembly with the ball seal, inlet sleeve, outlet seal, torsion spring, ball seal lift-off cams, two ball-to-housing bearings, two ball-to-shaft bearings, a thrust bearing, and two shaft seals. Metal parts of the valve are titanium and nickel- or iron-based alloys. The ball seal is machined Kel-F and the shaft seals are graphite-filled polyimide resin. Teflon bushings and sleeves are used throughout the valve to prevent metal-to-metal contact between moving metal parts. A drain is located between the shaft seals to route any leakage overboard. The fuel system purge port is located on the valve housing.

Initial motion in opening the valve is a 15-degree rotation of the shaft and cams to retract the lift-off seal. When the seal is retracted, the flow of fuel to the engine system starts. The ball and shaft interlock engages after the initial 15 degrees of rotation. At this time, the flow passages in the shaft and ball are aligned, and the shaft and ball rotate together to the commanded position. On closing, the ball moves with the shaft because of the load in the torsional spring and is stopped by a mechanical stop when the shaft is 15 degrees from the full closed position. The shaft and cams continue to rotate and reseat the ball seal during the final 15 degrees of travel. The ball is held closed by the preloaded torsional spring.

The MFV is a line replaceable unit (LRU) and weighs approximately 80 pounds.

304-128T

MAIN FUEL VALVE

Geometry

Titanium Housing, Inconel 718 Shaft, Ball and Bellows
440C Shaft Seal Seat

Flow Passage Diameter, inches	2.500
Ball Diameter, inches	5.125
Shaft Diameter (Seal), inches	1.875
Valve Length, inches	10.25

Operating Parameters (NPL MR = 6.0)

Inlet Pressure, psia	6104
Outlet Pressure, psia	6084
Temperature, R	93
Flowrate, lb/sec	144.6

304-128

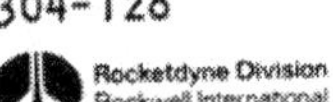

FUEL PREBURNER OXIDIZER VALVE

The fuel preburner oxidizer valve (FPOV) is a hydraulically actuated, spring-loaded-closed, ball-type valve with a 1.1-inch propellant flow passage. The FPOV is flange mounted between the oxidizer supply line to the fuel preburner (FPB) and the FPB oxidizer inlet. The valve permits or stops the flow of oxidizer to the FPB and the FBP augmented spark igniter (ASI). Valve position is controlled by commands from the engine controller; during mainstage operation the valve is modulated to control engine mixture ratio.

Major parts of the valve are a housing, shaft, ball, bellows assembly with the ball seal, inlet sleeve, outlet sleeve, torsion spring, ball seal lift-off cams, two cam-to-housing bearings, two ball-to-shaft bearings, a thrust bearing, and two shaft seals. Metal parts of the valve are nickel- or iron-based alloys. The ball seal is machined Kel-F and the shaft seals are graphite-filled polyimide resin. Teflon bushings and sleeves are used throughout the valve to prevent metal-to-metal contact between moving metal parts. A drain is located between the shaft seals to route any leakage overboard.

Initial motion in opening the valve is a 15-degree rotation of the shaft and cams to retract the lift-off seal. When the seal is retracted, the flow of oxidizer to the FPB starts. The ball and shaft interlock engages after the initial 15 degrees of rotation. At this time, the flow passages in the shaft and ball are aligned, and the shaft and ball rotate together to the commanded position. On closing, the ball moves with the shaft because of the load in the torsional spring and is stopped by a mechanical stop when the shaft is 15 degrees from the full closed position. The shaft and cams continue to rotate and reseat the ball seal during the final 15 degrees of travel. The ball is held closed by the preloaded torsional spring.

The FPOV is a line replaceable unit (LRU) and weighs approximately 35 pounds.

304-129T

FUEL PREBURNER OXIDIZER VALVE

Geometry	
Inconel 718 Housing, Shaft, Ball and Bellows 440C Shaft Seal Seat	
Flow Passage Diameter, inches	1.100
Ball Diameter, inches	2.968
Shaft Diameter (Seal), inches	1.100
Valve Length, inches	6.812
Operating Parameters (NPL MR = 6.0)	
Inlet Pressure, psia	7612
Outlet Pressure, psia	6050
Temperature, R	206
Flowrate, lb/sec	64.2

304-129

OXIDIZER PREBURNER OXIDIZER VALVE

The oxidizer preburner oxidizer valve (OPOV) is a hydraulically actuated, spring-loaded-closed, ball-type valve with a 1.1- by 0.285-inch propellant flow slot. The OPOV is flange mounted between the oxidizer supply line to the oxidizer preburner (OPB) and the OPB oxidizer inlet. The valve permits or stops the flow of oxidizer to the OPB and the OPB augmented spark igniter (ASI). Valve position is controlled by commands from the engine controller; during mainstage operation the valve is modulated to control engine thrust between minimum power level (MPL) and emergency power level (EPL).

Major parts of the valve are a housing, shaft, ball, bellows assembly with the ball seal, inlet sleeve, outlet sleeve, torsion spring, ball seal lift-off cams, two cam-to-housing bearings, two ball-to-shaft bearings, a thrust bearing, and two shaft seals. Metal parts of the valve are nickel- or iron-based alloys. The ball seal is machined Kel-F and the shaft seals are graphite-filled polyimide resin. Teflon bushings and sleeves are used throughout the valve to prevent metal-to-metal contact between moving metal parts. A drain is located between the shaft seals to route any leakage overboard.

Initial motion in opening the valve is a 15-degree rotation of the shaft and cams to retract the lift-off seal. When the seal is retracted, the flow of oxidizer to the OPB ASI starts. The ball and shaft interlock engages after the initial 15 degrees of rotation. At this time the flow passages in the shaft and ball are aligned, and the shaft and ball rotate together to the commanded position. On closing, the ball moves with the shaft because of the load in the torsional spring and is stopped by a mechanical stop when the shaft is 15 degrees from the full closed position. The shaft and cams continue to rotate and reseat the ball seal during the final 15 degrees of travel. The ball is held closed by the preloaded torsional spring.

The OPOV is a line replaceable unit (LRU) and weighs approximately 35 pounds.

304-130T

OXIDIZER PREBURNER OXIDIZER VALVE

Geometry	
Inconel 718 Housing, Shaft, Ball and Bellows 440C Shaft Seal Seat	
Flow Passage Dimensions, inches	1.100 X 0.285
Ball Diameter, inches	2.968
Shaft Diameter (Seal), inches	1.100
Valve Length, inches	6.812
Operating Parameters (NPL MR = 6.0)	
Inlet Pressure, psia	7586
Outlet Pressure, psia	5923
Temperature, R	206
Flowrate, lb/sec	21.3

CHAMBER COOLANT VALVE

The chamber coolant valve (CCV) is a hydraulically actuated, gate-type valve that serves as a throttling control to maintain proper fuel flow through the main combustion chamber and nozzle coolant circuits. The valve is installed in the chamber coolant valve duct which is an integral component of the nozzle forward manifold assembly and provides the housing for the valve.

Basically, the CCV consists of gate and cartridge assemblies. The gate assembly is supported by two roller bearings separated by a bearing spacer. One bearing interfaces with the chamber coolant valve duct and the other with the cartridge assembly. A third thrust bearing absorbs the axial loading on the valve gate. The gate contains a 1.6-inch-ID flow tube and is internally splined to a three-piece coupling that spline connects to the CCV actuator shaft. Rotary motion transmitted to the gate by the actuator shaft positions the gate flow tube passage relative to a 1.6-inch-ID flow passage in the bearing spacer to function as a throttle for the control of fuel flow through the valve. The CCV is not required to effect a positive shutoff and the 0.012-inch clearance that exists between the gate and spacer is not sealed.

The cartridge assembly is flange attached to the CCV duct and retains the gate assembly in the duct. It contains an indexing pin, six studs, and an external spline section for indexing and attaching the hydraulic actuator. A glass-fabric-epoxy-based thermal insulator minimizes heat transfer between actuator and cartridge to prevent freezing of the hydraulic fluid in the actuator. Doubly redundant shaft seals are used to seal the interface between gate shaft and cartridge and a drain is located between them to route any leakage overboard. A burst diaphragm, ported to the coupling cavity, precludes cavity rupture in the event the primary shaft seal fails.

The valve gate is positioned by commands from the engine controller. Its position is scheduled linearly with thrust reference so as to be 50 percent open at minimum power level (MPL) and full open at normal power level (NPL).

Metal parts of the valve are primarily nickel- or iron-based alloys. Shaft seals are graphite-filled polyimide resin. The valve is a line replaceable unit (LRU) weighing approximately 27 pounds.

Geometry	
Inconel 718 Housing, 21-6-9 CRES Gate Sleeve, A-286 CRES Gate Shaft and Rod, 440C Shaft Seal Seat	
Flow Passage Diameter, inches	1.600
Gate Diameter, inches	2.500
Shaft Diameter (Seal), inches	1.100
Valve Length, inches	7.729
Operating Parameters (NPL MR = 6.0)	
Inlet Pressure, psia	5995
Outlet Pressure, psia	5954
Temperature, R	93
Flowrate, lb/sec	75.7

304-131T

CHAMBER COOLANT CONTROL VALVE

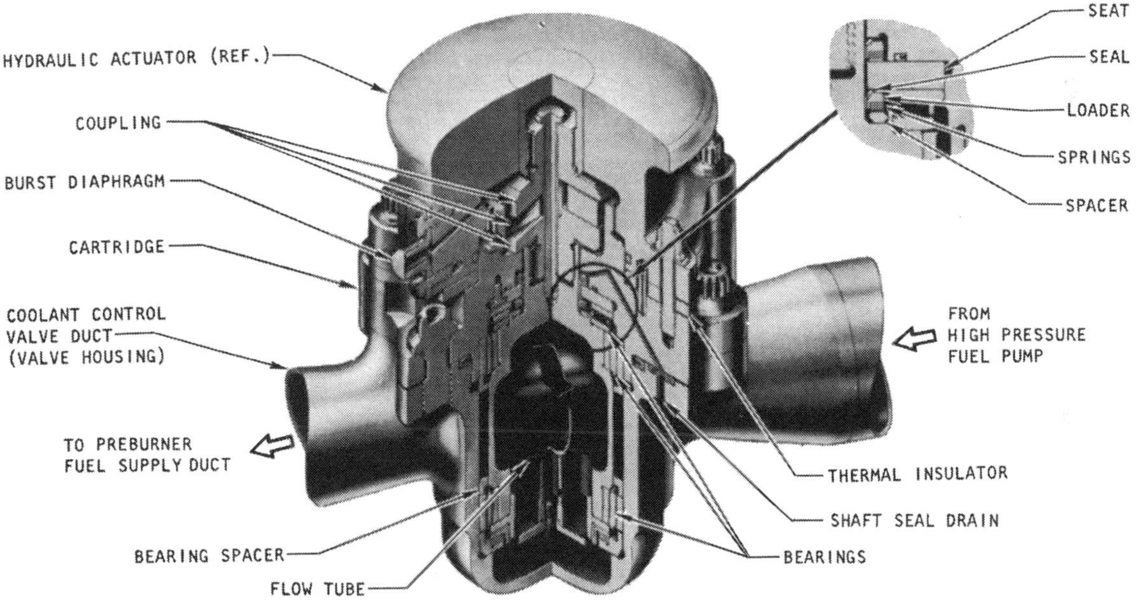

HYDRAULIC ACTUATOR (REF.)
COUPLING
BURST DIAPHRAGM
CARTRIDGE
COOLANT CONTROL VALVE DUCT (VALVE HOUSING)
TO PREBURNER FUEL SUPPLY DUCT
BEARING SPACER
FLOW TUBE
SEAT
SEAL
LOADER
SPRINGS
SPACER
FROM HIGH PRESSURE FUEL PUMP
THERMAL INSULATOR
SHAFT SEAL DRAIN
BEARINGS

304-131

PROPELLANT BLEED VALVES

The oxidizer bleed valve (OBV) and fuel bleed valve (FBV) are spring- and bellows-loaded-closed, pneumatically actuated-open, unbalanced area, metal-to-metal seat, poppet valves. The valves are opened by pneumatic pressure from the pneumatic control assembly (PCA) during engine start preparation to provide a recirculation flow for propellants through the engine to ensure that propellants in the engine are at the required temperatures for engine start. At engine start, the valves are closed by venting the actuation pressure. The valves are fail-safe in that they will close during engine start even if the actuation pressure is not vented. Fail-safe closing is accomplished by increasing system pressure acting on the unbalanced area poppet, combined with spring and bellows forces to overcome the actuation pressure. The valves have linear variable differential transformers (LVDT) for position indication.

The OBV inlet is flange mounted to the preburner oxidizer supply duct at the fuel preburner oxidizer valve (FPOV) location and the oxidizer bleed duct is welded to the valve outlet. The FBV inlet is flange mounted to the fuel high-pressure duct and the fuel bleed duct is welded to the valve outlet.

Major parts of the valves are a housing, poppet, poppet stem, bellows, spring, actuator cylinder, end cap, and position indicator. Nickel-base alloy was selected as the material for valve parts and Teflon sleeves and bushing are used between sliding parts to eliminate metal-to-metal interfaces. The valve assemblies, with their ducts, are line replaceable units (LRU) and the position indicators are LRU's, independent of the assemblies.

304-132T

PROPELLANT BLEED VALVE

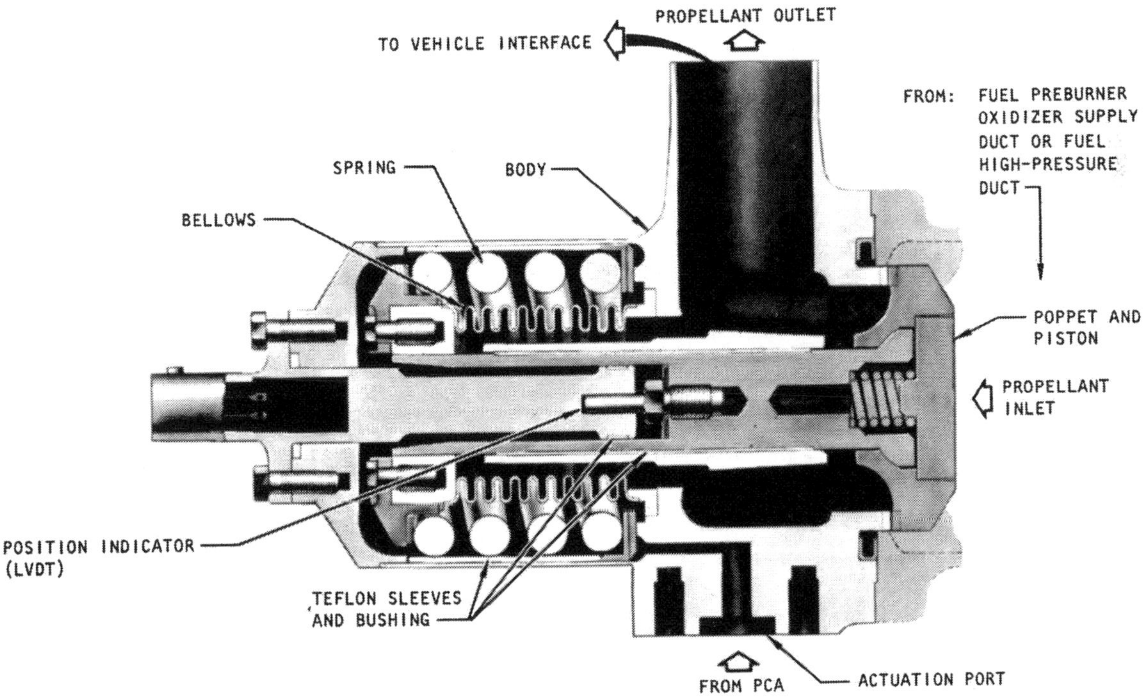

PROPELLANT OUTLET

TO VEHICLE INTERFACE

FROM: FUEL PREBURNER OXIDIZER SUPPLY DUCT OR FUEL HIGH-PRESSURE DUCT

SPRING

BODY

BELLOWS

POPPET AND PISTON

PROPELLANT INLET

POSITION INDICATOR (LVDT)

TEFLON SLEEVES AND BUSHING

FROM PCA

ACTUATION PORT

304-132

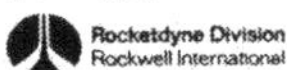

ANTIFLOOD VALVE

The antiflood valve (AFV) is a spring- and bellows-loaded, normally closed, poppet-type valve. It is flange attached to the preburner oxidizer pump supply duct and prevents liquid oxygen from prematurely entering the heat exchanger and causing vehicle oxidizer tank ullage pressure collapse at engine start. The AFV opens during engine start when oxidizer pressure reaches approximately 250 psid. It remains open during engine operation and reseats closed during engine shutdown when oxidizer pressure decays to approximately 200 psid. The bellows cavity is ported to a burst diaphragm that is set to rupture at approximately 60 psig, referencing the cavity to atmosphere, if the bellows should crack or leak.

The tungsten carbide poppet pivots so that the sealing face of the poppet aligns itself with the sealing face of the seat when the sealing faces contact during valve closing.

Self-generating contamination is prevented by using Teflon or Rulon bushings and sleeves to eliminate metal-to-metal rubbing and sliding contact.

304-133T

HEAT EXCHANGER ANTI-FLOOD VALVE

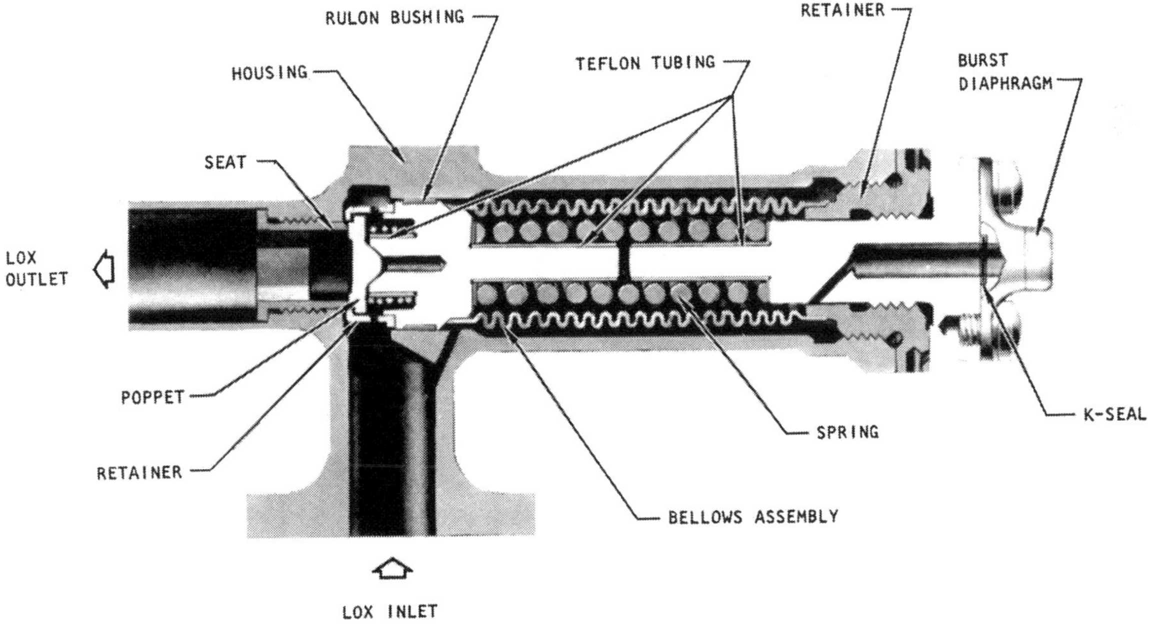

RULON BUSHING
RETAINER
HOUSING
TEFLON TUBING
BURST DIAPHRAGM
SEAT
LOX OUTLET
POPPET
RETAINER
K-SEAL
SPRING
BELLOWS ASSEMBLY
LOX INLET

304-133

PURGE CHECK VALVES

The pneumatic control system purge check valves are spring-loaded, normally closed, pressure-actuated-open, poppet valves that isolate propellants from the pneumatic system.

Five valves of two different configurations are used to purge the fuel system, the oxidizer domes and igniters (3), and the high-pressure oxidizer turbopump turbine seal. The two configurations differ in mounting flange hole pattern (8 or 7 holes), and flowrates as controlled by two sizes of valve body bore diameters (0.8088 or 0.8298) in which the poppet seats. The valve cap is welded to the valve body and to the end of the purge line. The valve body is flange attached to the system.

Valves crack open when inlet pressure reaches 20 psig and snaps to the full open position (preventing valve chatter) when the inlet pressure is applied to the full poppet diameter.

A flat-lapped metal-to-metal seal prevents reverse leakage through the valve, and improves service and storage life.

The valve housing and poppet are Haynes 188 and the spring is 302 CRES. Machined Teflon spring guides and heat-shrinkable Teflon tubing around the poppet stem eliminate galling of metal parts and self-generating contamination.

304-134T

PURGE CHECK VALVE

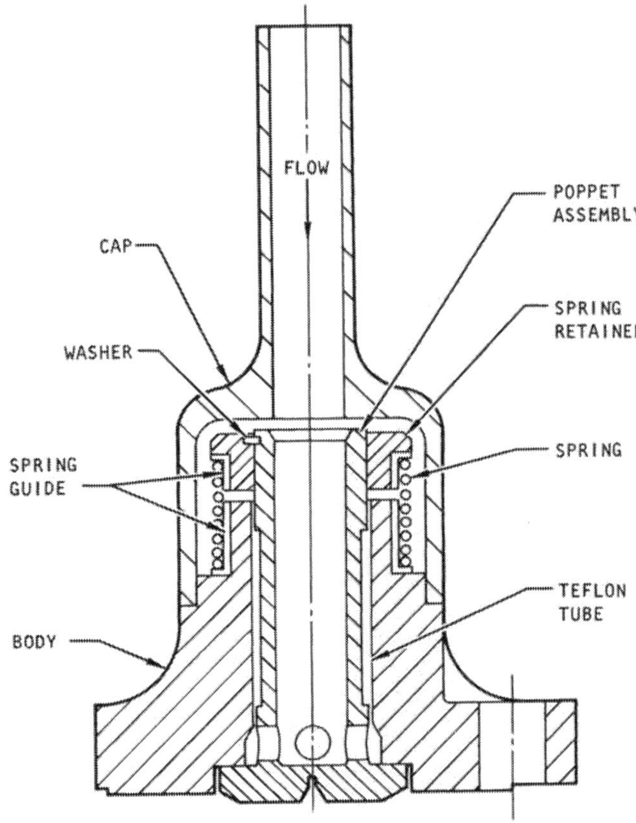

304-134

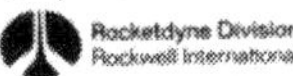

CONTROLLER

The controller is a pressurized and thermally conditioned electronics package attached to the thrust chamber and nozzle coolant outlet manifolds on the low-pressure fuel turbopump side of the engine. It is designed to operate in conjunction with engine sensors and the vehicle control system for engine control, monitoring, and checkout operations. The controller receives engine performance data from the engine sensors and control commands and requests for data from the vehicle through the vehicle engine electronics interface (VEEI). It also transmits engine data to the vehicle through the VEEI.

Controller features include:

1. Built-in self-test that enables it to perform continuous self-tests in flight and on the ground
2. Component checkout (individually or group) on the ground upon command from the vehicle
3. Recycle the engine to a ready condition with 5 minutes after a launch abort without any ground servicing
4. Changeable computer go/no-go limits (all or partial) by inserting a modified memory program

The controller is divided into five functional sections arranged on a dual-redundant basis:

1. Input Electronics—Receives data from the in-flight sensors, converts the data to a digital form, and sends it to the computer
2. Computer Interface Electronics—Controls the flow of all data within the controller
3. Digital Computer Unit—Performs computations and issues engine control signals upon receipt of sensor data and vehicle commands, and stores engine data until requested by the vehicle; conducts tests of all control system components once every 20 milliseconds
4. Output Electronics—Converts computer digital commands to voltages suitable for operating engine igniter, actuators, and on/off controls
5. Power Supply Electronics—Converts vehicle-supplied electrical power to voltages required by controller functional units

304-135T

CONTROLLER ASSEMBLY

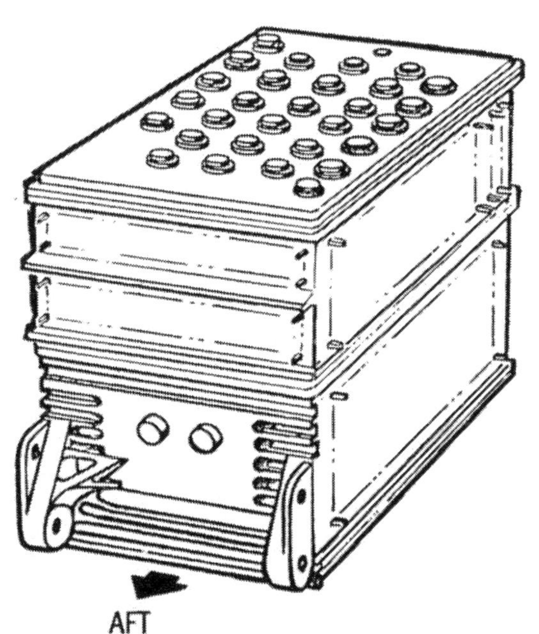

- ELECTRICAL COMPUTER CONTROLLER OF ENGINE
- WEIGHT - 167 POUNDS
- SIZE - 14.5 X 17.0 X 23.5 INCHES
- TEMPERATURE

  -50 F TO +95 F OPERATING
  -200 F TO +200 F NONOPERATING
- POWER - 615 WATTS
- CONNECTORS

  22 ENGINE
  6 GSE

304-135

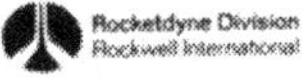

IGNITION SYSTEM

The ignition system is an augmented spark igniter (ASI) type that initiates combustion of main propellants in the main combustion chamber (MCC) and in the fuel preburner and oxidizer preburner (FPB and OPB).

The system consists of six inductive-discharge spark igniters, which have the spark exciter electronic circuitry integrally packaged with the spark plug, and three ASI combustion chambers, which are integral with the injector and dome assemblies of the main chamber and preburners. Ignition system propellants supplied to each of the ASI chambers are ignited by dual-redundant spark igniters that are controlled and supplied with 28-vdc power by separate power sources from the engine controller. The ASI combustion chamber hot gases discharge into the MCC and preburner combustion chambers to ignite the main propellants. The ignition system produces a continuous supply of hot gas throughout the engine duty cycle, although the electrical energy source is removed within 3.5 seconds after engine start.

OPERATING CHARACTERISTICS (NPL MR = 6.0)

	Main Chamber	OPB	FPB
Oxidizer Flowrate, lb/sec	0.67	0.69	0.76
Fuel Flowrate, lb/sec	0.94	1.30	1.12
Mixture Ratio (o/f)	0.71	0.53	0.67

304-136T

IGNITION SYSTEM

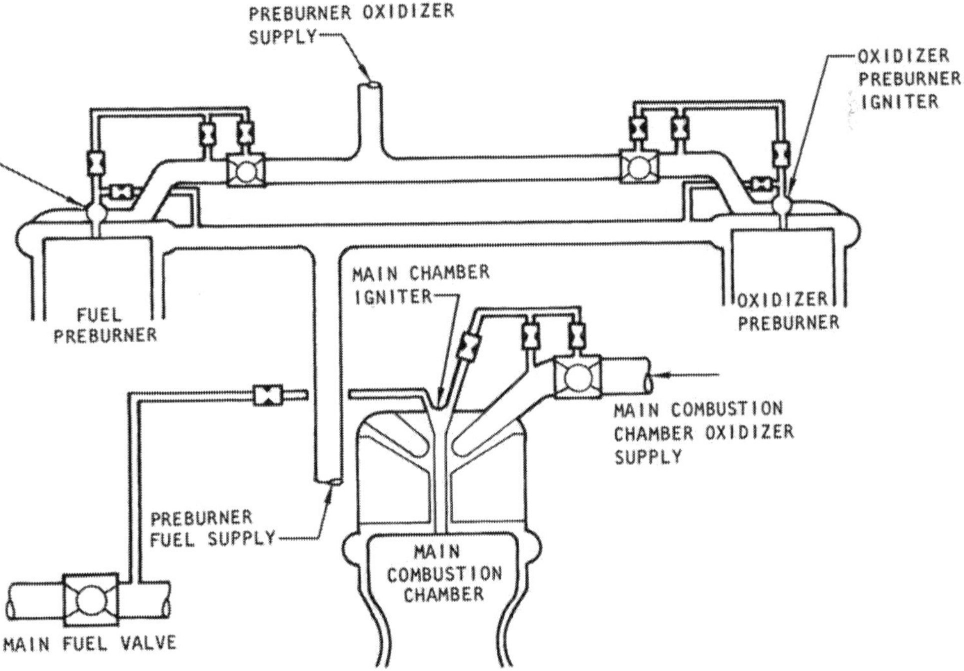

PREBURNER OXIDIZER SUPPLY

FUEL PREBURNER IGNITER

OXIDIZER PREBURNER IGNITER

MAIN CHAMBER IGNITER

FUEL PREBURNER

OXIDIZER PREBURNER

MAIN COMBUSTION CHAMBER OXIDIZER SUPPLY

PREBURNER FUEL SUPPLY

MAIN COMBUSTION CHAMBER

MAIN FUEL VALVE

304-136

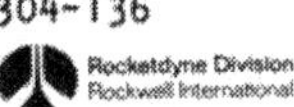

ASI INJECTOR/COMBUSTION CHAMBER

The augmented spark igniter (ASI) injector and combustion chamber augment the spark igniters in establishing ignition of the main propellants for the main and preburner combustion chambers.

Both injectors and chambers are integral parts of the main chamber and preburner injector assemblies. The oxidizer injector consists of two orifices, 180 degrees apart, that direct the ASI oxidizer into the centerline of the ASI chamber at the plane of the spark igniter electrodes.

Eight orifices comprise the fuel injector, directing the ASI fuel tangentially into the ASI combustion chamber at a plane approximately 0.5 inch downstream of the oxidizer orifices. This injection flow pattern creates an oxidizer-rich condition in the vicinity of the spark igniter electrodes prior to ignition. After ignition, the flow pattern develops into a relatively low mixture ratio environment around the spark igniter electrodes and an oxidizer-rich core along the combustion chamber length surrounded by a fuel-rich zone that protects the ASI combustion chamber walls.

304-137T

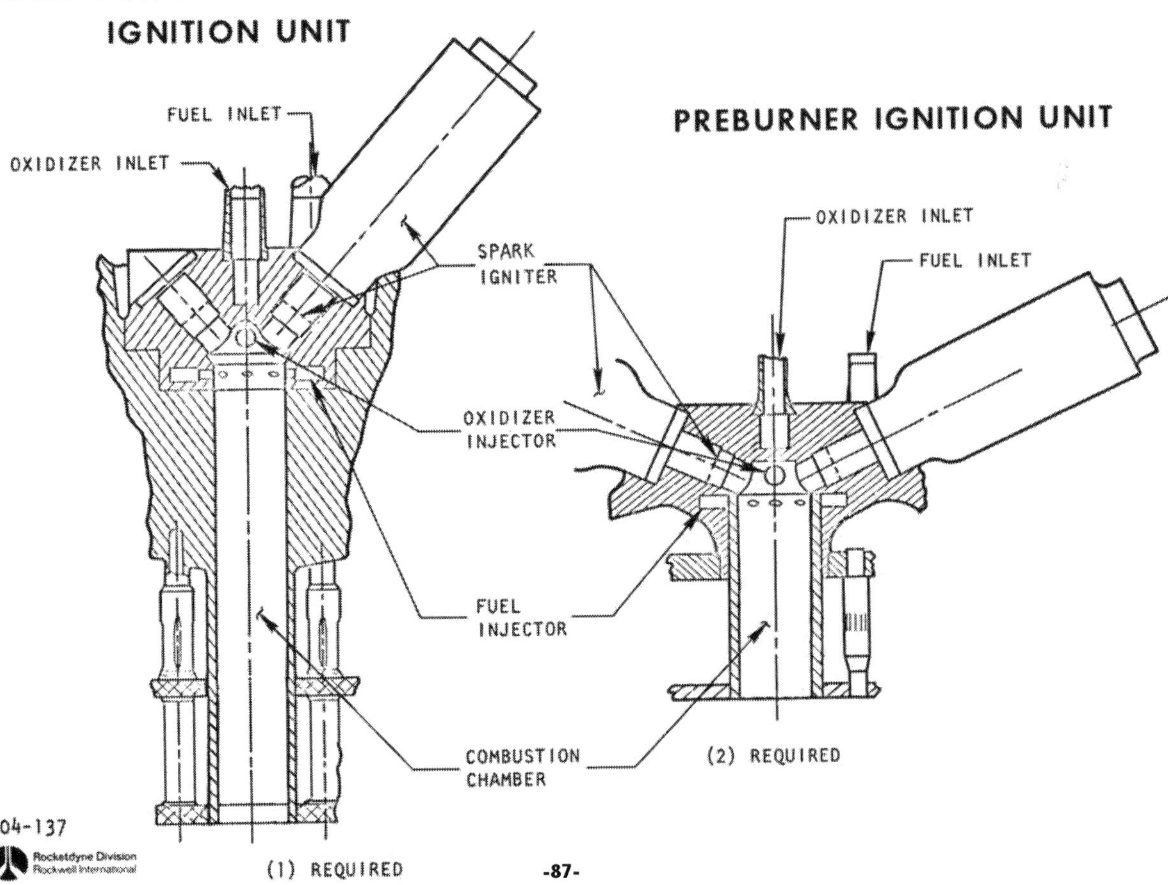
MAIN COMBUSTION CHAMBER
IGNITION UNIT
FUEL INLET
OXIDIZER INLET
SPARK
IGNITER
OXIDIZER
INJECTOR
FUEL
INJECTOR
COMBUSTION
CHAMBER
(1) REQUIRED
PREBURNER IGNITION UNIT
OXIDIZER INLET
FUEL INLET
(2) REQUIRED
304-137
Rocketdyne Division
Rockwell International

SPARK IGNITER

The spark igniter is a dual-series-gap, high voltage, sparking device that ignites the propellants supplied to the ignition system. The spark igniter combines a spark plug and spark exciter electronics in an integral, hermetically sealed unit that transforms a low-level 28-vdc input into a 10-kilovolt output at a minimum of 50 sparks a second.

The spark igniter is approximately 5.3 inches long and 2.0 inches at its maximum diameter. It weighs approximately 1.6 pounds.

The solid-state exciter electronics are encapsulated in rigid potting within the igniter housing, and the final steps of assembly include pressurizing the inside of the housing with nitrogen containing a 10-percent helium tracer, to provide a hermetically sealed unit impervious to moisture.

The dual-surface spark gaps are in series and are formed by the center electrode and intermediate electrode (0.025-inch gap), and by the intermediate electrode and outer electrode (0.025-inch gap).

The igniters are installed into threaded ports of the ASI chambers. A captive Naflex seal, retained by a spring retainer that ensures positive seal placement during installation, seals the interface of igniter and ASI chamber.

304-138T

INTEGRAL SPARK IGNITER

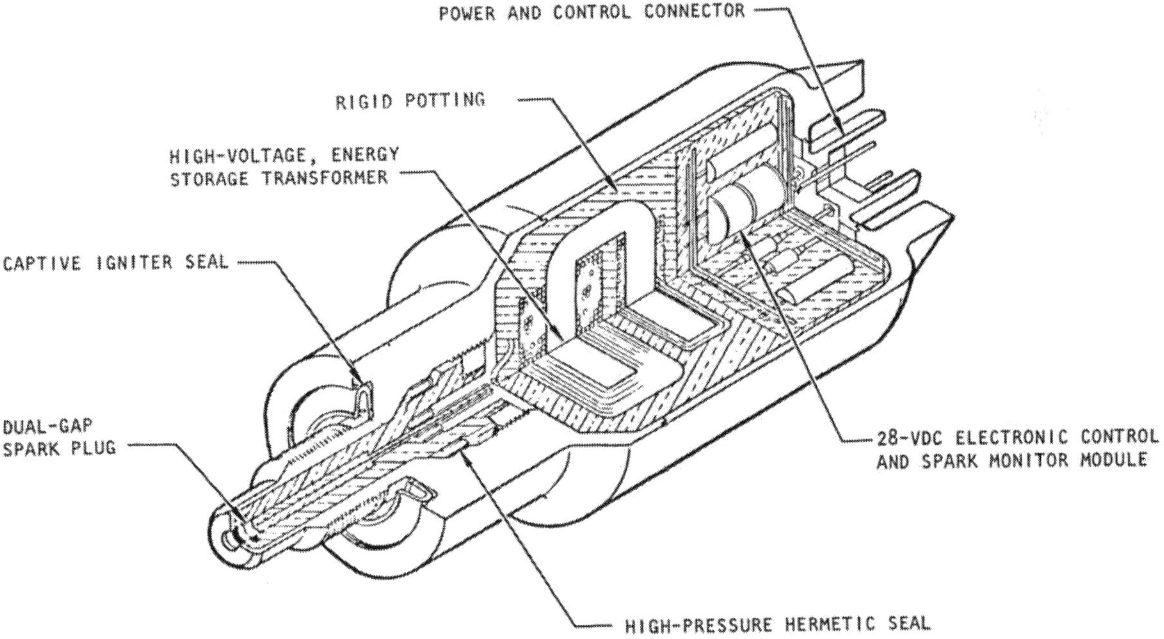

POWER AND CONTROL CONNECTOR

RIGID POTTING

HIGH-VOLTAGE, ENERGY
STORAGE TRANSFORMER

CAPTIVE IGNITER SEAL

DUAL-GAP
SPARK PLUG

28-VDC ELECTRONIC CONTROL
AND SPARK MONITOR MODULE

HIGH-PRESSURE HERMETIC SEAL

304-138

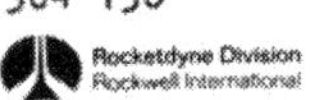

GIMBAL BEARING

The gimbal bearing assembly is a spherical, low friction, universal joint that has ball and socket bearing surfaces. The bearing assembly provides the mechanical interface with the vehicle for transmitting thrust loads and permits angulation of the actual thrust vector about each of two vector control axes. The gimbal bearing is attached to the engine main injector by ten bolts through off-set bushings that allow lateral positioning of the bearing. The gimbal bearing position is established by optical alignment during engine buildup to ensure that the actual thrust vector is within 30 minutes of arc to the engine centerline and 0.60 inches of the gimbal center. Cyclic life is obtained by low friction, anti-galling, bearing surfaces that operate under high loads. A Teflon fiberglass material is bonded on all surfaces that transmit engine thrust and dry gimbaling loads. A permanent, resin-bonded, dry film lubricant is used on the torsional bearing surfaces. Major parts of the titanium alloy assembly are a body, anti-torque shaft, anti-torque block, and a seat. The assembly has an overall dimensional envelope of approximately 11.3 by 14 inches and weighs approximately 104 pounds.

304-139T

GIMBAL BEARING ASSEMBLY

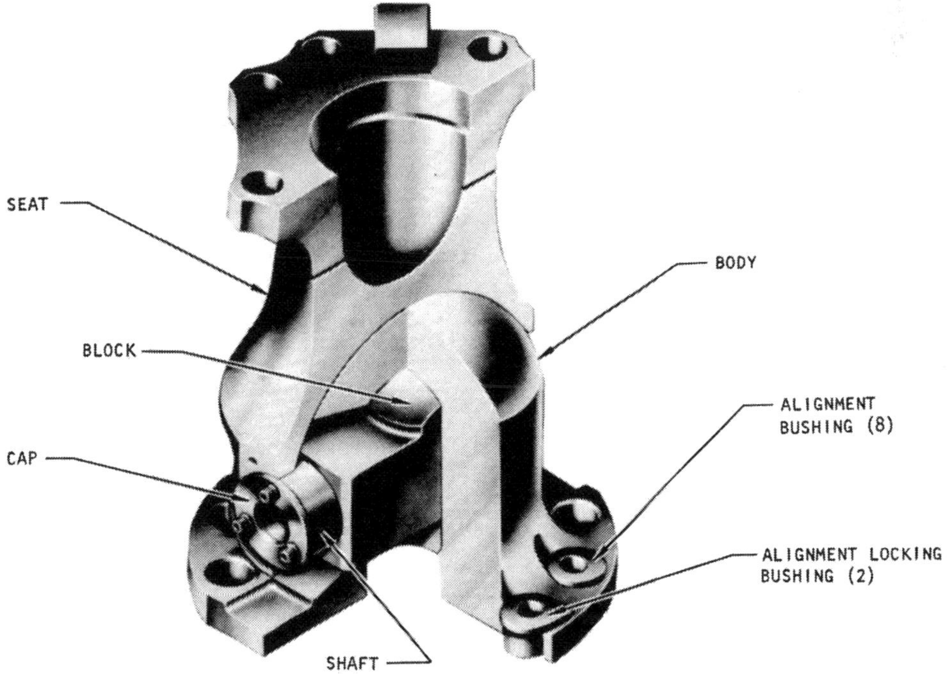

SEAT

BODY

BLOCK

ALIGNMENT BUSHING (8)

CAP

ALIGNMENT LOCKING BUSHING (2)

SHAFT

304-139

ISBN #978-1-937684-79-2
www.PeriscopeFilm.com

Made in the USA
Lexington, KY
30 March 2016